C

S

Volume One

Matthew J. Hatami, P.E.

OILFIELD SURVIVAL GUIDE

V O L U M E   O N E

Tactics | Procedures | Checklists | Fatality Analysis

Short Stories | Train Wrecks | Court Cases

Mass Disasters | Troubleshooting

Problem Job Prevention

Life Saving Skills

Oilfield Books, LLC
Practical Oil & Gas Publications
Oklahoma City, Oklahoma

First published 2016 by Oilfield Books, LLC

Oilfield Survival Guide, Volume One
Premier volume in the Oilfield Survival Series of books

Other than *Piper Alpha* and *Deepwater Horizon*, the names of incidents, oil and gas reservoirs, companies, and people involved in the situations contained within this book have been changed to protect the privacy of businesses and individuals. Any resemblance to actual persons, locations, or companies is purely coincidental.

Published by Oilfield Books, LLC
Oklahoma City, Oklahoma USA

Oilfield Survival™ is a trademark of Oilfield Books, LLC

For discounts on bulk purchases e-mail
oilfieldsurvivalguide@yahoo.com

Follow Oilfield Survival Guide on: Facebook @OilfieldSurvivalGuide
Twitter @OilfieldGuide, Instagram @OilfieldSurvivalGuide

Concept, format, and design by Matthew J. Hatami
Cover design by Matthew J. Hatami

ISBN 978-0-692-81308-9

First Edition: November 2016

Dedicated
to Those who
lost their life,
their mind,
their finances
in this business
of oil and risk

WARNING

Never apply any techniques, core tactics, procedures, checklists or anything written or quoted in this book without performing thorough calculations, hazard/risk assessments, statistical analysis, and due diligence with your company, team, and everyone involved with your operation.

Accordingly, the publisher and author cannot accept any responsibility for anything that occurs as a result of the use or misuse of any techniques, tactics, procedures, checklists or anything written or quoted in this book. As stated in the book, never take a procedure, tactic, recommendation, suggestion, or anything else and jam it into use in your area of operations. This is dangerous and you are going to have problems.

Table of Contents

Oilfield Survival Series

OILFIELD SURVIVAL GUIDE

VOLUME I

FOR ALL OILFIELD SITUATIONS

Preface

OILFIELD SURVIVAL

Mistakes are painful in the oil and gas business, both financially and physically. Many have been wiped out and forgotten. A statistic in a chart somewhere gathering dust. A distant memory, if remembered at all. However, each has a story to tell. A story of those final moments of life and death. A story that just might save your skin, if you're interested and willing to listen. Within the confusion and chaos of oilfield disasters are valuable lessons from the honorable efforts of great explorers. Great oil men and women dealing with adversity, fighting to survive. This book gives tribute to them.

The cause of undesirable oilfield situations is sometimes simple, sometimes complex, but always unforgiving. Often, it comes down to one's actions or inaction. Hesitate to shut a job down for safety, men die. Poor communication on location, men die. Neglect to perform the proper calculations, men die. Lack of attention to detail, men die. When the stakes are high, there is little room for error, as will be shown throughout the book when we step into the steel toe boots of others as they deal with oilfield challenges. Learning from the experience of others is a key principle of the Oilfield Survival Series. The philosophy is best summed up with a 2,200-year-old quote from old Latin literature:

"Te de aliis, quam alios de te suaviust." – Plautus [1]

Translation

"It is better to learn from the mistakes of others than that others should learn from you." [2]

I

MANAGING OILFIELD RISK

High cost operations, below type-curve performance, equipment failure, troubled jobs, train wrecks, property damage, near-misses, environmental incidents, accidents, and fatalities do not just happen. They are mostly caused by people's actions or inaction. They are the undesirable result of human error. And they are preventable.

Based on analysis of over 1,000 undesirable oilfield situations reviewed to create the Oilfield Survival Series of books, key issues in the majority of oilfield "situations" include technical and operational proficiency, communication, situational awareness, distraction, and fatigue, with the root cause emanating from a lack of knowledge, intelligence (ability to apply knowledge), and attention to detail.

Obtaining well-rounded oil and gas experience takes time and can be expensive. Catastrophic mistakes made in the process can cost hundreds of millions and even billions of dollars. Those numbers do not include injury or loss of life – concepts that are beyond the human ability to truly quantify, even though the courts and lawyers work hard to do so.

"Risk comes from not knowing what you're doing."

Warren Buffett, Billionaire Investor[1]

In the oil business, you are given two barrels: one barrel is for luck, the other is for experience. This book will help fill the barrel with experience before the barrel of luck runs out. Learn from the mistakes of others. Stand on the shoulders of giants – it's less costly than learning the hard way. Unfortunately, for certain types of people the hard way is the only way. Ultimately, there is always a price to pay.

As you will see in *Volume One*, the pain of oilfield mistakes is delivered in more ways than one. The Oilfield Survival Series was created to help manage risk and avoid pain through the conveyance of knowledge: clear, direct and to the point, no non-sense and no waste. Time is the most valuable commodity we have, even more valuable than oil or money, for that matter.

OVER 1,000 INCIDENTS
Reviewed and incorporated to formulate the foundation of knowledge included in the Oilfield Survival Guide

Although this book is designed to help increase your oil and gas knowledge and intelligence in managing risk to prevent undesirable oilfield situations, there is no substitute for real world experience. When something expensive and painful occurs, as you are living through the turmoil and details of it, the memory is burned into your brain with such power that it is impossible to forget or ignore.

Additionally, real world experience is highly motivational, forcing the situation to be analyzed from every angle, not only in corporate meetings, safety meetings, investigations, and the courts, but also in your mind. People tend to play out unfortunate situations over and over again in their mind so that if a similar situation occurs in the future, the appropriate steps and actions are taken to manage the risk and prevent a negative outcome.

"I have known no wise people who didn't read all the time – none, zero."
Charlie Munger, Billionaire Investor[2]

To convey a broad spectrum of real world experience and knowledge, the Oilfield Survival Series and premier book, "Oilfield Survival Guide" (OSG), employs a number of methods to increase your practical knowledge and intelligence in this business of oil and risk. These methods include the following:

Core Tactics

OSG stands on a foundation of Core Oilfield Tactics – clear, direct, short and to the point guidelines to reduce risk and instill principles of oilfield survival. Core Tactics address things you should or should not do, whether you work in the office or in the field, are a top executive in a suit and tie, or a new hand first day on the job in jeans and a t-shirt, no steel toes yet. In Volume One, 50 Core Tactics are presented, subdivided into groups of 10.

Core Tactics 1 through 50 are listed in a certain order. Not in order of importance, each tactic is important in its own way. Core Tactics are presented in an optimal order to convey experience through the knowledge- and intelligence-increasing process as you read the book. Volume One starts with broad foundational Core Tactics and builds upon them, progressing to different areas of the oil and gas business. Do not dismiss or skip reading certain tactics because you think they are too simplistic. Operations in this business quickly become complex, as illustrated with examples embedded in short stories, fatality reports, and court cases accompanying each Core Tactic. Reading detailed examples of other people making mistakes will help instill the foundational principles of oilfield survival.

The largest oil & gas property damage loss, in terms of property value, is the Piper Alpha oil production platform explosion, July 6th 1988, with a loss value of $1.81 billion (2013 USD). As a comparison, the Deepwater Horizon drilling rig explosion, April 20th 2010, had a property damage loss of ~$600 million (for the loss of the rig) not including the billions in spill related liability claims.

Marsh, The 100 Largest Losses[3]

Procedures

Written procedures are critical in this business. Clear procedures significantly reduce risk and prevent problems. In Chapter 5, a step-by-step process to create a variety of oilfield procedures is presented. Additionally, a number of example procedures are provided as a reference tool and guide to help construct your own procedures.

Oilfield procedures must be customized and tailored for a given job and area of operation. For example, taking a procedure from the Marcellus and using it in the Permian is not recommended. There will be key differences, including State regulations, equipment availability, operating environment, operator and service company policies, subsurface characteristics, people on location, and experience level. The list goes on and is addressed in greater detail later in the book.

Some folks do not believe in written procedures and do not use them. Most of the time, these folks have problem jobs, are high cost operators, and/or have suboptimal production performance. In this business, there is far too much to remember to not have it written down. Additionally, as humans, we do not operate at the same level throughout the work day. People get tired, distracted, stressed, panicked, scared, excited and every other emotion you can think of. All of these things impact our ability to perform. There is only so much we can remember at any given time. The bottom-line is that we, as humans, forget stuff. Procedures help ensure we remember, operate safely, and are as efficient as possible.

Safety, environment, performance, and cost are critical in the oil business and must be managed through good process. On the flipside, some operators go overboard. Some engineers have procedures that are so long, convoluted, and complex – they are almost impossible to follow. Additionally, in many cases, the procedures are out of touch with actual field operations.

"A fool with a plan can outsmart a genius with no plan."

T. Boone Pickens, Billionaire Oilman[4]

Procedures must always be written in conjunction with field hands, service company personnel, and operating company representatives. At the very least, oilfield procedures must be "signed off" by the folks that are going to implement the plan on location. Procedures must be reviewed thoroughly and approved by all. Never drop a new procedure on someone and go on vacation or disappear. If you do, you are a fool.

Procedures do not have to be long and bureaucratic. They can be as simple as a one-paragraph email. You want to keep them clear and simple. They need to be functional and useful. I have seen a one-page procedure accomplish more good than a ten-page procedure full of every little operational detail. The ten-page procedure did not focus on the key issues and areas that usually have the most problems. The ten-page procedure ended up increasing cost and causing problems because it was a distraction from what mattered most.

However, for certain operations you must include a lot of detail. Although the procedure will be long, it will be successful if you construct it correctly and design it to focus on potential issues and critical tasks. Finally, procedures should be viewed as a work in progress, continually updated and refined to address areas that need attention. Procedures are a key aspect of oilfield survival and will be addressed in great detail throughout the book.

Checklists

Written checklists have significantly reduced risk in a variety of complex endeavors. In the OSG, oil and gas operations are frequently compared to aviation and space exploration operations as there are a number of similarities, especially when it comes to managing uncertainty and complexity. Additionally, aircraft incidents and accidents are very well documented and researched, providing valuable information for cross-industry knowledge transfer.

"My best friend is a person who will give me a book I have not read."

Abraham Lincoln, U.S. President[5]

Regarding checklists, the Federal Aviation Administration (FAA) requires pilots to use written checklists when operating aircraft. A number of well-known and catastrophic aircraft accidents have been caused by not using checklists. The most well-known incident is the crash of Northwest Airlines Flight 255 shortly after takeoff in August 1987, in which 148 passengers and 6 crewmembers were killed. Only 1 passenger, a 4-year-old girl, survived. Additionally, since the plane crashed onto a busy roadway striking vehicles, 2 people on the ground were killed.[6]

Based on the National Transportation Safety Board (NTSB) accident report, "the probable cause of the accident was the flight crew's failure to use the TAXI CHECKLIST to ensure that the flaps and slats were extended for takeoff."[7] Following the investigation, the U.S. Department of Transportation and FAA issued a report titled *"The Use and Design of Flight Crew Checklists and Manuals,"* finding that from 1983 to 1986, there were 21 accidents/incidents of aircraft that occurred due to checklist issues. In five cases, a checklist was not used at all.[8] The Aviation Safety Reporting System (ASRS), a confidential voluntary safety reporting system used by frontline aviation personnel, had 195 reports of situations involving checklists during the preceding five-year period. In 43% of ASRS reports, 84 situations, the flight crew did not use a checklist or missed key items on the checklist.[9] Based on these statistics and similar oilfield situations we will review in the book, it is clear that checklists hold significant value.

Checklists help ensure smooth, predictable, and efficient operations. If you are working with a complex oilfield operation, you definitely want to construct a checklist to cover key points. If you are given a long convoluted procedure, consider boiling it down to core items and build a sequential checklist.

"Every 7 seconds a worker is injured on the job."

National Safety Council[10]

Engineering Design

Design flaws and technical mistakes can set an operation up to fail before it begins. A number of oilfield situations are reviewed in which poor design and weak engineering contributed to major issues during field implementation. There are several key techniques that can help ensure strong design quality, reduce the odds of engineering mistakes, minimize operational cost, and maximize production performance. These techniques are reviewed throughout the book.

Operations Analysis

Reviewing oilfield operations and providing a project post-mortem, highlighting key learnings, practical knowledge, useful tips and best practices is a key aspect of the Oilfield Survival Series. The goal is to extract usable information and tools to help you avoid troubled jobs and train wrecks. For those that are unfamiliar with the term "train wreck," in the oilfield, a train wreck is a severely troubled operation.

Often, there are small details or warning signs that must be investigated. For example, have you ever noticed a difference in pressure readings on the same line, a small drip leak at the wellhead, the bit not acting right, an unexpected pressure or change in pressure, pulling more or less weight than expected, a piece of equipment that does not look right or sound as it should, a person that seems unfocused or unaware during critical operations? These are all warning signs of potential train wrecks that can be avoided if you have the knowledge, are focused on the details, and take action.

TACTICAL TIP

Learning to listen to your gut instincts can be a powerful tool in avoiding oilfield train wrecks.

If something seems 'off,' consider shutting the job down and stepping out of the chaos of the situation, even if just for a few minutes, to reflect on what is happening.

"Face reality as it is, not as it was or as you wish it to be."

Jack Welch, fmr. CEO GE[11]

Short Stories

Stories have been used for thousands of years to help pass knowledge down from generation to generation. Stories are an excellent tool in the knowledge-building process because they make information come alive. For example, if someone suggests using an 80% safety factor, it comes across as plain information. Just a rule some back-office manager cooked up. The chance it will be correctly followed is low.

However, if someone suggests using a safety factor and tells multiple stories about people who did not use a safety factor, or did not use a safety factor correctly, and provides detailed examples about the devastation that occurred, you are more likely to remember to include safety factors in your operation and apply them correctly for the environment you are working in.

> ***TACTICAL TIP***
>
> Operating without safety factors can be hazardous to your happiness, health, and wealth. Make it your business to know the safety factors, even if they are zero. At least you know before you go down for the dirt nap.

The stories in this book are based on a variety of situations across the industry, spud to plug. The stories may or may not be true. Details may be changed from actual events, certain aspects may be embellished or understated. Names and locations are always removed or changed to protect the associated. Although there is no substitute for actual experience, Oilfield Survival stories come as close as possible to the real thing by including a lot of detailed and graphic information to help you remember them. Since the point of this book is to ensure you don't make the same mistakes for yourself, you have to remember the information when it counts.

If an operations boss wants oilfield hands to remember safety procedures and techniques, they should utilize emotional storytelling.

Safety Beyond Compliance[12]

Fatality and Accident Reports

Whenever people are hurt or killed on the job, it is a tragedy, and I have the deepest sympathy, compassion, and respect for the family and co-workers. Most workplace deaths can be prevented. Oilfield accidents and fatalities are rarely acts of God. Almost all cases are due to human error. When someone dies on the job, investigations occur, the government gets involved, courts and lawyers often become involved, and the situation gets a lot of attention, as it should. From an informational point of view, for the purposes of learning, there is significant value.

Fatality reports are utilized in the OSG as a very effective learning tool. When someone is killed on the job, and all the factors leading up to the incident are presented, with the horrible consequences as a result, it gets into your head. This ensures you do not forget when it counts, in the middle of an operation, when there are a dozen distractions and you are thinking about everything else except how you, or someone on your team, can get killed.

Fatalities are a sensitive subject. Some folks may think it should not be discussed out of respect for the deceased. My view is that it is out of respect that we discuss it, and not ignore or pretend that it did not happen. If we sweep fatalities under the rug and learn nothing from them, what is to prevent the tragedy from happening again and again?

Consider this: a situation unfolds right before your eyes, and you remember a similar occurrence from the OSG; then you remember the fatality associated with the situation. You take action, saving your life, the lives of others, and saving your company millions of dollars. The person who was killed in the Fatality Report and from whom you learned, saved your life, or someone else's life, with their life and their story.

"We don't keep secrets and cover things up. We do it all up front and in public. That's the way freedom is, and we wouldn't change it for a minute."

Ronald Reagan, U.S. President[13]

Court Cases

The legal process generates a significant amount of analysis, commentary, and associated documentation. Resolution of a dispute in a court of law can take a long time, a decade in some cases, with a significant amount of paperwork being generated throughout the entire process. Due to the substantial material value associated with many of the situations that involve the legal system, I extracted the most valuable and useful information from oil and gas related court cases to help increase your knowledge and operating ability in this business. You need to know how oilfield situations end up in court. Do not think that you are immune from ending up in court or with legal action pressed against you. Even if you primarily push a pencil around a piece of paper all day, you can still end up in court and potentially in prison, for something that happened on location that you had nominal involvement with.

Knowledge Box

Towards the bottom and on the side of various pages, a "knowledge box" is provided for fast knowledge in the form of statistics, quotes, facts, warnings, and oilfield survival pearls of wisdom. Packed with information, backed in many cases with quantitative research, the knowledge boxes are an excellent quick reference.

"We had vision, saw the vast possibilities of the oil industry, stood at the center of it, and brought our knowledge and imagination and business experience to bear in a dozen - 20, 30 directions. There was no branch of the business in which we did not make money."

John D. Rockefeller Sr.,
Billionaire
Oilman[14]

II

CORE TACTICS: 1 - 10

Mistakes can be as deadly as a bullet in this business. Core Tactics, the foundational principles of oilfield survival designed to tactically address critical aspects of oil and gas operations in the field and office, were developed to help eliminate mistakes.

Core Tactics 1 through 10 start at the highest level and quickly get into detail. You will learn attention to detail is key in minimizing oilfield risk. If you learn nothing else from this book, you will learn the value of attention to detail. You will learn it from the mistakes of others, from the pain and suffering of others, I promise you.

When reading through OSG situations, take the time to consider how it could happen to you. No one is invincible or immune from making fatal mistakes. There are far too many variables in play to predict the future. With uncertainty, there is risk, and risk must be managed and minimized. Nothing worth doing in the oilfield is without effort. Remember, safety is hard work.

"S-H-K Drilling Company had a hard and fast rule: when stuck, try everything, except to simply try to pull the pipe loose. It rarely worked and might get you fired. This had been set up years before by both Skinny Hunter and Dick Hodges. It was an excellent rule, had saved thousands of dollars, and kept many tool pushers out of difficulty."

Gerald Lynch, Toolpusher[1]

1. ALWAYS: Protect Yourself and Your Company

The mistakes of just one person can destroy an entire company. When hired into the oil and gas industry, due to the nature of the business, a tremendous amount of responsibility is given. When much is given, much is expected. You should be honored to work in one of the greatest industries of our time. Arguably, there is no other business that provides the opportunity to work on hundred-million-dollar projects starting on day one.

PROTECT YOURSELF
AND YOUR COMPANY
PRIORITY PLEDGE

1. SAFETY
2. ENVIRONMENT
3. JOB *

** NEVER HESITATE*
to shut down an unsafe job

With the rise of horizontal shale multi-well pad sites, there is easily over $50 million US dollars of assets on a single location in which you are currently working. Within the course of the first year in the oilfield family, a person will easily touch over $1.0 billion dollars in value at work. At this point, one may think, "They don't pay me enough!" Well, my friends, the oil and gas business is known to, and has a reputation for, providing excellent salaries for those with knowledge. Gaining experience is key: the more you know, the more you can deliver and the more money you will make. This book will help; protecting your company provides the foundation.

Just like in football, a strong defense is key. Protecting your company is the foundation of a strong oilfield defensive strategy. Your company provides for you and gives not only financial compensation, benefits, and hopefully days off, but also the opportunity to gain experience. You should want to protect your company.

Property value of 100 largest oil & gas property damage losses is >$34 billion, not including well control costs or 3rd party liability claims

Marsh, The 100 Largest Losses[2]

If you are a disgruntled employee, unhappy at work, or hate your company, you should quit as soon as possible. Find a place to work that makes you happy, or get out of this business. A person with a negative attitude is a dangerous person, not only to himself but to all those around. If you work with a disgruntled person, watch out. A perpetually angry person that hates the company and oilfield work is not focused, not interested in doing a great job, and not interested in protecting the company.

> ***FACT***
>
> Nearly 13,000 American workers are injured on the job each day.
>
> *National Safety Council*[3]

Your company also wants to protect you. No oil and gas entity wants to harm the employees or cause undesirable oilfield situations. It's incredibly expensive and not good business. If you feel that this is not the case with the company you work for, my guess is that it is not the company as a whole, but individual people and/or the customer who do not know what they are doing. They most likely lack knowledge, intelligence, and attention to detail.

If you determine that careless coworkers, poor management or dangerous customers are an issue, you have several options:

A. Transfer jobs or locations and move far away from the risk.
B. Quit the company. Learn from the experience and find a better working environment.
C. Become a disgruntled employee.
D. Put yourself, co-workers and company at risk by "just doing it." Ignore the risk and hope nothing happens. As they say, no one has been killed yet – *famous last words and not what the OSG suggests.*
E. Take action, find a solution, address the risk, and move forward safely – *OSG's preferred choice and what Core Tactics are designed for.*

"Your work is going to fill a large part of your life, and the only way to be truly satisfied is to do what you believe is great work. And the only way to do great work is to love what you do."

Steve Jobs, Apple[4]

Never hesitate to stop an unsafe job. If you identify a safety issue, or a person doing anything unsafe, address it immediately. In many cases, exercising Stop Work Authority (SWA) by stopping dangerous activity and correcting that activity to perform the work safely only takes a few minutes. Core Tactic #2, *NEVER: Hesitate to Shut Down an Unsafe Job*, illustrates this foundational oilfield survival principle with examples.

For some situations, you may find it difficult to shut the job down. Although it is rare, there can be certain situations in which there will be push-back when you decide to shut a job down for safety reasons. You may need to convince another person to shut the job down. An aggressive customer/client or boss may disagree with your opinion on what is too risky or dangerous. A more experienced coworker may dismiss your concerns. If you are challenged, do not back down. Protect yourself and your company. Consider the Plan of Action step-by-step tactical approach below which you can implement over the course of a few minutes or a few days, depending on the situation and urgency required:

1) Confirm the source of the problem is one or more risky colleagues, managers, or customers (company man, representative, engineer etc.). Identify exactly who is pushing the dangerous situation or not taking action. If you find it's not a person but a process or procedure, fix it and write up a new procedure – writing procedures is covered in detail with examples in Chapter 5. Once you confirm it's an actual person or group of people, go to Step 2.

2) Explain the issues, concerns, and risks in as much detail as possible to the person or people pushing the high-risk approach. Have a face-to-face conversation, if possible, and systematically address everything in an organized fashion. Try not to involve emotion because it can be taken personally. Additionally, if the person pushing has a strong personality,

they may disagree with you as a power move, for no other reason than to show they are the boss. Remember, the rough and tough nature of the oil business attracts certain personalities that must be dealt with tactfully.

If the situation is unfolding on location and someone is asking or telling you to do something that you feel is unsafe, under no circumstances should you do it. However, don't curse out the customer or boss and say that what he is asking is stupid and that you would never do it. Use a little sophistication and consider saying something like this:

"I have a few questions, concerns, not sure I understand how to best manage the risk involved, let's talk about it for a moment, interested to hear your thoughts."

Or more simply:

"Before I do this, let's discuss the risks."

The customer, boss, or co-worker might respond:

"What risks? I don't see any. Now do your job!"

Your response:

"Well, I see X, Y, and Z as potential risks."

The goal is to initiate a conversation to address the risk and minimize it as much as possible. It's your life that's on the line. Ask the customer, boss, or co-worker to explain how they would address the risk. Ask to see the written procedures, Job Safety Analysis (JSA), Material Safety Data Sheets (MSDS), equipment documents, and/or calculations so you can study them. Can you watch the person that is pushing the unsafe act do it, so you can see how they manage the risk?

---

823 oil and gas workers were killed on the job, from 2003 to 2010, a fatality rate seven times greater than the rate for all U.S. industries.

OSHA, United States Department of Labor[5]

Just because someone with more experience does something does not make it right or safe. If it looks wrong to you, trust your gut instinct. If you ask the person who is pushing the dangerous work to consider doing the job themselves and they hesitate even for a few seconds with a concerned look, you just confirmed that there is unmanaged risk that must be addressed.

Under no circumstances should you let the work go on. In my experience, just by suggesting they do the work will get the issues solved. Make sure you tell them all the issues you see and get them to help address and manage the risks. The act of suggesting the high risk person consider the dangerous action themselves is to get them to hesitate, think, and address the risk. In most cases, "Mr. High-Risk" will not perform the high risk operation, which is why he is asking you to do it. If this backfires and "Mr. High-Risk" decides to attempt the dangerous action or tell someone else to do it, IMMEDIATELY MOVE TO STEP 10.

3) If you cannot have a face-to-face discussion or phone call, send a well-organized email or text message, addressing all concerns. In many cases, a text or email is better because the written word can be more impactful. Send it only to the person pushing, don't CC everyone yet. Let the person consider your concerns.

4) Discuss and offer solutions and options, never ultimatums. Any good solution is better than no solution. If you only bring objections, problems, and no solutions, your leverage will be significantly reduced during the discussion, and the probability that your concerns will be dismissed or overruled is high. If your concerns are dismissed and solutions ignored, move to Step 5.

"Maintain 'stop work' policies that establish the responsibility and authority for all employees and contractors to stop work they believe to be unsafe."

Ryan Lance, CEO ConocoPhillips[6]

5) Now is the time to loop in others and/or cc others to get additional perspectives. If there is someone with more experience than you who also sees the risk, get their opinion and potential solutions.

6) If you work on the oilfield services side of the business and are dealing with an unsafe customer, loop in your boss, other advisors, and engineers, informing them of the situation. If it's your boss who is pushing a dangerous situation without addressing the risk properly, consider writing out your concerns and loop in a few advisors on your level. Don't go over your supervisor's head just yet. Always be respectful. At this point, it's important to get everything into writing and potential solutions into writing so everyone knows you're serious and concerned. Also, an email correspondence allows multiple people to provide suggestions and slows down the operation, allowing people to think. If there is no time and each minute counts, utilize group text messaging, which is generally considered more urgent. Group text messaging is very effective when everyone is engaged. If none of this works and the situation escalates without progress, go to Step 7.

7) Get EHS (Environment, Health and Safety) personnel involved by phone, text, email, and get them on location immediately.

8) Share examples, where a similar situation resulted in an undesirable outcome. This book is full of actual oilfield situations that you can use to make a point. If none of this works and the person or group decides to move forward without addressing the risk, go to Step 9.

9) With your superiors engaged and involved, systemically go through all of your potential solutions and options to get the

job done safely. If this does not work, and none of your solutions are accepted, move to Step 10.

10) **SHUT THE JOB DOWN** – Stop everything and make a point of it. You must shut the job down and protect your company and your customer, even if it's your boss or client who unknowingly puts people at risk. Once you shut the job down and have EHS involved, the situation generally will get the exposure and attention to address the risk in question. Almost all companies have a written policy that any person on location can shut a job down. It is common knowledge in the oilfield. Most companies mention this at safety meetings every day and if they don't, they should, as a constant reminder. It's part of the foundation of oilfield safety. Everyone in this business has the authority, right, and obligation to stop dangerous situations and shut the job down. This empowers anyone and everyone to protect the company.

"Protect Your Company" refers to much more than the Company as an entity or the owners and shareholders. The 1st Law of Oilfield Survival "Protect Yourself and Your Company" includes protecting:

- Teammates / Coworkers
- The Public / Local Residents
- Customer /Clients
- Environment
- Land Owners
- Mineral Owners
- Supervisors / Executive Leadership
- Company Reputation
- Assets
- Oil and Gas Industry

"We reward people who exercise stop-work authority, to underscore both its importance and our aim to confront any problem right then and there."

John Watson, CEO Chevron[7]

You must protect everything and everyone with good judgement and a solid work ethic. Safety is hard work, and attention to detail is critical. Do not let the little things slip past you. Discipline is paramount as mistakes can be unforgiving in this business. It only takes one time. You can be the best and safest worker 364 days of the year, and one day let your guard down, lack attention to detail, and you and your company may pay a great price.

"For this reason the gentleman will:

1) Employ a man on a distant mission and observe his degree of loyalty.

2) Employ him close at hand and observe his degree of respect.

3) Hand him troublesome affairs and observe how well he manages them.

4) Suddenly ask his advice and observe how wisely he answers.

5) Exact some difficult promise from him and see how well he keeps it.

6) Turn over funds to him and see with what benevolence he dispenses them.

7) Inform him of the danger he is in and note how faithful he is to his duties.

8) Get him drunk with wine and observe how well he handles himself.

9) Place him in mixed company and see what effect beauty has upon him.

By applying these 9 tests, you may determine who is the unworthy man."

Confucius, Teacher[8]

Deepwater Horizon Disaster

Eleven people were killed and 4.9 million barrels of oil were spilled (initial rate of 62,000 bopd declining to 53,000 bopd over 87 days)[9] in what is considered one of the largest oil spills in human history.[10] The disaster was the result of a loss of well control after cementing operations were thought to be successfully concluded. In 2015, BP agreed to pay $18.7 billion to the U.S. government and 5 states to settle all federal and state claims.[11] This is considered the largest corporate settlement in history.[12] Total cost of the accident is estimated at over $60 billion.

Mistakes Cost Money[13]

Costs and Claims Category	Amount (USD)
BP Claims, Advances, and Settlements	$14,730,000,000
BP Spill Response and Cleanup	$14,000,000,000
BP Natural Resource Damages	$7,300,000,000
BP Economic Claims to Five States	$5,900,000,000
BP Clean Water Act Penalty	$5,500,000,000
BP Costs Associated with the Spill	$5,200,000,000
BP Federal Criminal Settlement	$4,000,000,000
Anadarko Settlement	$4,000,000,000
Transocean Civil and Criminal Fines	$1,400,000,000
BP Damage Assessment Process	$1,300,000,000
Halliburton Settlement	$1,100,000,000
MOEX Settlement	$1,065,000,000
BP Local Government Claims	$1,000,000,000
Deepwater Horizon Rig Loss	$600,000,000
Cameron Settlement	$250,000,000
Weatherford Settlement	$75,000,000
Est. Total Pre-Tax Cost of Accident	**$67,420,000,000**

"If I have seen further it is by standing on the shoulders of giants."

Sir Isaac Newton, Scientist[14]

Let's walk through the events leading up to the Deepwater Horizon blowout and perform a deep-dive analysis, including minute-by-minute account of operations on April 20, 2010.

Disaster Key Facts[15]

- **Deepwater Horizon - Semisubmersible drilling rig**
    - Dynamically positioned.
    - Owned by Transocean.
    - Under lease to BP since 2001.
    - Arrived on Block 252, January 31, 2010.
- **Mississippi Canyon Block 252 lease**
    - 9 square mile plot in Gulf of Mexico.
    - $34 million payment to U.S. Mineral Management Service (MMS) for the lease.
    - BP operator; non-op partners Anadarko and MOEX.
    - Geology uncertain, based on available 3-D seismic and offset well data, thought to be a potential commercial discovery.
        - Mid-Miocene age turbidite sands primary objective.
- **Macondo Well**
    - 1st well on Block 252 lease.
    - ML-mudline @ 5,067 ft (sea water depth).
    - Planned true vertical depth (TVD) of 20,600 ft below sea level.
    - $96 million drilling budget.
    - Scheduled for 51 days of work.
    - Daily op cost $1,000,000 (incl. fuel, expendables, services etc.)
    - Marianas semisubmersible drilling rig began work on Macondo in October 2009.
        - Drilled for 34 days reaching 9,090 ft before stopping due to Hurricane Ida.
        - Marianas was damaged by the Hurricane and replaced by Deepwater Horizon.

- Multiple significant difficulties occurred during drilling.
    - Kick (fluid influx) at 8,970' detected, shut-in well, mud weight increased, circulated kick out.
    - Kick at 13,305' detected, shut-in well, pipe got stuck, severed pipe at 12,146', sidetracked well.
    - Ballooning at 15,113', lost fluids, mud weight decreased, then fluid influx.
    - 8 significant lost circulation events; total loss 16,000 bbls of mud costing $13 million in time and materials.

Disaster Timeline[16]

Factual information in the timeline is reproduced from the National Commission's reports, court records, trial documents, witness testimony, U.S. Coast Guard reports, and the BP investigation. There is conflicting information regarding certain events. Below is my best effort at constructing a timeline of events leading up to the disaster.

April 9, 2010

- At 18,193 ft lost circulation occurs. 172 bbl pill is pumped.
- Pill successful, circulation regained, mud returning to surface.
- BP engineering determines 14.0 pound per gallon (ppg) mud necessary to balance pressure exerted by the reservoir.
- Calculations indicate drilling with 14.0 ppg mud would yield a 14.5 ppg equivalent circulating density (ECD). A 0.5 ppg increase in density risks further fracturing of the reservoir rock, leading to additional lost circulation events. Drilling continues to 18,360 ft.
- BP engineering concludes they have run out of drilling margin, inform partners Anadarko and MOEX that "well integrity and safety" issues require them to stop drilling.

April 11 to April 16, 2010

- Macondo well is stable and balanced.

- Open hole logging commences for 5 days.
- BP concludes reservoir has sufficient size and pressure to be economic. Decision made to install production casing.
    - Reservoir estimated at 50 MMbbls.
- Preparations made for casing and cementing operations; several aspects make cementing Macondo challenging.
    - Lost circulation history during drilling.
    - Narrow drilling margin: range between pore and frac pressure.
        - 14.7 ppg frac gradient.
        - 14.0 ppg pore pressure gradient.
- Original plan is to run a long string design; casing from total depth to the mudline.
    - BP engineers, working with Halliburton, simulate the likely outcome of long string design with various cement strategies.
    - Results indicate long string cannot be cemented reliably.
    - Team switches to liner design.
    - There is resistance, within BP, to switch.
    - BP asks Halliburton to run numerous simulations in effort to find a way to make the long string casing a viable option.
        - Team engages BP cementing expert to review Halliburton's recommendation.
            - BP expert determines various simulation model inputs must be corrected. Calculations with the new inputs show that a long string could be cemented properly.

> ***TEST YOUR SKILLS***
>
> **What is a potential issue with the approach employed to solve the long string vs. liner dilemma?**
>
> *They appear to approach the problem by trying to find a way to make a long string work instead of asking what design option will best address the cementing difficulties they face.*[17]

TEST YOUR SKILLS: LONG STRING VS. LINER

Why can it be harder to cement a long string versus a liner?

1) *Cementing a long string typically requires higher pumping pressures compared to a liner; therefore, you may have a higher probability of fracturing the formation during pumping.*[18]

   Due to higher pumping pressures with the chosen long string design on Macondo, engineers made other changes to the cement job to reduce ECD including:
   A. *Reducing the volume of cement pumped*
   B. *Reducing the pump rate*
   C. *Employing foam cement (nitrogen + cement)*[19]

2) *Increased risk of cement contamination because the cement travels across a larger surface area, leading to increased exposure to mud and cuttings that adhere to the casing. Risk of cement contamination is increased on Macondo because the long string is tapered, making it harder for wiper plugs to clean the casing.*[20]

3) *Liner hanger includes a mechanical seal adding a barrier to annular flow; therefore, a successful cement job is not as critical compared to a long string.*[21]

4) *With a liner you can remediate a bad cement job without having to rely on the cement as a barrier to flow.*[22]

For the Macondo well, what is the benefit of a long string design?

Offers better long term well integrity compared to a liner tieback, as long as you have a good cement job.[23]

- BP decides to run the long string design and a total of 51 bbls of cement (60 bbls after foaming):
  - 48 bbl Class H foam cement
  - 12 bbl Class H cement

- Design calls for 16 centralizers. Weatherford only has 6 centralizer subs in stock, designed to screw into place between joints of casing.
  - Alternative is slip-on centralizers (installed on casing exterior).
  - BP Team Leader, distrusts slip-on centralizers with separate stop collars because they can slide out of position or hang up on other equipment as casing is run.
  - Halliburton runs simulations using OptiCem™ program to predict if mud channeling will occur. Calculations suggest casing needs more than 6 centralizers.
  - Halliburton tells BP about the problem on April 15th. BP Drilling Engineering Leader obtains permission from a senior manager to order 15 additional slip-on centralizers, the most BP can transport immediately by helicopter.
  - That evening, Halliburton preforms additional simulations, finding channeling due to gas flow would be less severe with 21 centralizers.
  - BP Engineer sends an e-mail to Team Leader, saying Halliburton performed additional modeling with the final directional surveys, caliper log, and 6 centralizers. "ECD at the base of the sand jumped up to 15.06 ppg. . . The model runs for 20 centralizers reduce ECD to 14.65 ppg which is back below the 14.7+ ECD we had when we lost circulation. . . we need to honor the modeling to be consistent with our previous decisions to go with the long string."

> ***TEST YOUR SKILLS***
>
> **Why is centralization especially important for this particular cement job and well?**
>
> *Low volume of cement (60 bbls) will be pumped on a long string design on a well with severe gas flow potential.*
>
> *Little room for error.*[24]

- o BP Team Leader learns the next day of the decision to add more centralizers: he initially defers, then challenges the decision.
- o When the centralizers arrive, BP Drilling Engineer out on the rig reports that the centralizers are conventional design with separate stop collars. He e-mails another drilling engineer to question the need for additional centralizers. Engineer responds that the team would "probably be fine" even without the additional centralizers and that "He is right on the risk/reward equation."
- o BP Team Leader complains, "Also it will take 10 hrs to install them. We are adding 45 pieces that can come off as a last minute addition. I do not like this and as it was approved in my absence I did not question but now I am very concerned about using them." In the end, BP Team Leader's view prevails; BP installs 6 centralizer subs.

- Before running the casing, the crew circulates bottoms up. No losses occur. The mud from the bottom of the well is inspected; it contains 1,120 gas units on a 3,000 unit scale (mud was sitting in place for 6 days). After additional circulation, gas decreased to 20 to 30 units.

April 18, 2010

- Crew runs long string (bottom to top):
  1) Reamer shoe @ 18,304 ft
  2) 189 ft of 7" 32# HCQ-125 casing
  3) Float collar: 2 flapper float valves, spaced one after the other, held open by an 'auto-fill tube'

> ***TEST YOUR SKILLS***
>
> **Are floats considered a reliable hydrocarbon flow barrier?**
>
> *No. They are not considered a barrier to hydrocarbon flow, they are designed to prevent cement backflow. API does not include float equipment among its list of subsurface mechanical barriers.*[25]

- Allows mud to pass through as casing is lowered into well
- Float valves convert from 2-way valves into 1-way valves once rate reaches 6 bpm, causing a differential pressure on the tube of 750 psi. The conversion allows flow down the casing but prevents fluid from coming back up the casing.

4) 5,627 ft of 7" 32# HCQ-125 casing
5) 7" to 9-5/8" XO @ 12,488 ft
6) 7,430 ft of 9-5/8" 62.8# Q-125 casing

April 19, 2010

- Completed long string run to 18,304 ft (took 37 hours).
- Pumps initiated but casing pressures up to 1,800 psi without establishing circulation.
- Day-shift Company Man (1st time on Deepwater Horizon, serving 4 days as a relief man), and the Drilling Engineer call Team Leader, their supervisor onshore. Speaking with him and Weatherford staff, the team decides to increase pump pressure in increments, hoping eventually to close the floats by dislodging the auto-fill tube.
  - Weatherford tells BP they could increase pressure up to 6,800 psi, but at 1,300 psi the ball would pass through the bottom of the auto-fill tube without converting/closing the floats.
  - On 9th attempt, pump pressure peaks at 3,142 psi and then suddenly drops, they start circulating mud.
  - Rate never exceeds 4.3 bpm.
  - Drilling mud subcontractor, M-I SWACO, predicted it would take 570 psi to circulate mud after activating the float

TEST YOUR SKILLS

Why did the float valves probably not close?

Auto-fill tube is designed to dislodge/convert in response to a 6 bpm flow-induced pressure. The tube is not designed to dislodge due to an increase then sudden drop in static pressure.[26]

valves. Instead, circulation pressure is 340 psi. The Day-shift Company Man expresses concern about it.

- He and the Transocean crew switch circulating pumps to see if that makes a difference.
- They conclude the pressure gauge is broken.

- Due to lost circulation concerns, BP makes several decisions in regards to the cement job:
  1) Circulate 350 bbls of mud before cementing, rather than 2,760 bbls needed for bottoms up circulation. BP fears the longer they circulate, the greater the risk of lost returns.
  2) Pump at 4 bpm or less.
  3) Limit volume of cement pumped to 60 bbls.
  4) BP determines that the annular cement column should extend 500' above the uppermost hydrocarbon zone, sufficient to fulfill MMS regulations of "500 feet above the uppermost hydrocarbon-bearing zone." However, this does not satisfy BP's own internal guidelines, which specify that the top of the annular cement should be 1,000 feet above the uppermost hydrocarbon zone.
  5) BP selects foam cement to lighten the resulting slurry from 16.7 ppg to 14.5 ppg
     - Cement stability tests run in February show the slurry was not stable. If the slurry does not remain stable up until the cement cures, the small nitrogen bubbles may coalesce into larger bubbles, rendering the hardened cement porous and permeable. If the instability is severe, nitrogen

TEST YOUR SKILLS

What are the benefits of bottoms up circulation?

- *Cleans well*
- *Reduces channeling*
- *Conditions mud*
- *Allows examination of drilling mud for hydrocarbon influx*

can break out of the cement. There was no evidence BP examined the test results sent by Halliburton.

- With more accurate temperature and pressure information in April, Halliburton conducts additional testing, including a foam stability test which shows the cement was once again unstable. Halliburton runs a 2nd stability test, increasing the pre-test "conditioning time" to 3 hours, which indicates the foam cement would be stable. This test begins at 2:00 a.m. on April 18th. The test takes 48 hours.

April 20, 2010

12:38 a.m. - Halliburton finishes pumping cement job. Full fluid returns are observed. A valve at the cementing unit is opened to check if floats are holding. Model predicts 5.0 bbls of flowback. Actual flowback was 5.5 bbls, tapering off to a trickle. Drill pipe pressure went from 1,150 psi to 0 psi. No flow observed. They conclude floats are holding.

Cement Job Schedule

Fluid	Volume (bbls)	Density (ppg)
Base Oil	7	6.7
Spacer	72	14.3
Unfoamed Lead Cement (Class H)	5	16.7
Foamed Cement (Class H)	39 (Foamed to 48)	14.5 BH Density
Unfoamed Tail Cement (Class H)	7	16.7
Spacer	20	14.3
Mud Displacement	857	14.1

"Cementing is 80% placement, and placement is 80% centralization. If you can't centralize the pipe and place cement in more than 99% of the annular areas, you can't expect to achieve reliable long-term isolation."

Mike Cowan, Senior Research Fellow[27]

TEST YOUR SKILLS: CEMENTING

Why are full returns and lift pressure an imperfect indicator of a successful cement job?

Positive signs of a successful cement job are:

1) Fluid flow-in equals flow-out, determined by accurate measurement, not determined solely by visual inspection.
2) Actual lift pressure matching modelled and calculated lift pressures.
3) Plug bumped on time.

***HOWEVER**, volume and pressure indicators cannot account for cement quality, channeling, and the exact location of the cement.*[28]

What are potential reasons for a cement job failure?

1) *Cement u-tubes due to failed floats*
2) *Mud contaminates the cement*
3) *Shoe track cement swaps places with rathole mud*
4) *Cement mixed incorrectly at the bulk plant or on location*
5) *Foam instability leads to nitrogen breakout*
6) *Lack of casing centralization causes cement channeling*
7) *Formation is fractured during cement job*
8) *Inadequate mud conditioning*
9) *Poor hole cleaning*
10) *Hydrocarbon influx*
11) *Poor engineering design*

1:00 a.m. - Crew begins process to install seal assembly in subsea wellhead. Two pressure tests are successful. Drill pipe pulled out of riser. (6 hours total)

5:45 a.m. - Halliburton Cement Field Engineer sends email to Account Rep, "We have completed the job and it went well." He attaches detailed report stating job was "pumped as planned" and they had full returns.

7:30 a.m. - Morning Meeting: BP discusses good news that the final cement job had gone fine. BP decides to skip the cement bond log (CBL), saving time and $128,000.

8:52 a.m. - BP Engineer e-mails the Houston office: "Just wanted to let everyone know the cement job went well. Pressures stayed low, but we had full returns on the entire job. . . We should be coming out of the hole shortly."

10:14 a.m. - BP's Drilling Operations Manager in charge of Macondo emails: "Great job guys!"

10:43 a.m. - BP Engineer, about to leave the rig on the helicopter with the Schlumberger team, sends an e-mail laying out his plan for conducting the well integrity test and subsequent temporary abandonment procedures. Few had seen the plan's details when the rig supervisors and members of the drill team gathered for the rig's daily 11 a.m. pre-tour meeting. It was the first time BP Company Men on the rig had seen the procedures they would use that day.

> ***TEST YOUR SKILLS***
>
> **What is the risk of dropping a new procedure on the crew at the last minute?**
>
> *Inadequate time for the crew to review, discuss, identify issues, address concerns, and prepare before implementation.*

Below is the procedure:

1) Test casing per APD to 250 / 2,500 psi
2) RIH 8,367'
3) Displace to seawater from there to above the wellhead

4) With seawater in the kill close annular and do a negative test ~2,350 psi differential
5) Open annular and continue displacement
6) Set a 300' balanced cement plug w/ 5 bbls in DP
7) POOH ~100-200' above top of cement and drop neft ball / circulate DS volume
8) Spot corrosion inhibitor in the open hole
9) POOH to just below the wellhead or above with the 3 ½" stinger (if desired wash with the 3 ½" / do not rotate / a separate run will not be made to wash as the displacement will clean up the wellhead)
10) POOH and make LIT / LDS runs
11) Test casing to 1,000 psi with seawater (non MMS test / BP DWOP) - surface plug
    a. Confirm bbls to pressure up on original casing test vs. bbls to test surface plug (should be less due to volume differences and fluid compressibility - seawater vs sobm)
    b. Plot on chart / send to Houston for confirmation

TEST YOUR SKILLS: WELL CONTROL

In terms of well control, what is the biggest risk in the procedure?

The primary flow barrier (drilling mud) will be removed, leaving the recently pumped cement as the primary hydrocarbon barrier in the well.

12:00 p.m. - Positive pressure test: blind shear ram (BSR) is closed. Pressure is increased to 250 psi for 5 minutes, then increased to 2,500 psi for 30 minutes. Pressure remains steady. The drilling crew and Day-shift Company Man consider the test successful. In preparation for the negative pressure test, crew runs drill pipe to 8,367 ft.

Spacer: BP tells M-I SWACO mud engineers to create a spacer out of two different lost-circulation materials left over on the rig. M-I SWACO had previously mixed two different lost circulation pills in case there were further lost returns. At BP's direction, M-I SWACO combined the lost circulation materials to create a large volume of spacer (424 bbls) never previously used by anyone, or tested to be used as a spacer.

1:28 p.m. - Mud starts to be offloaded from Deepwater Horizon to supply vessel M/V Damon Bankston.

2:30 p.m. - Helicopter lands, four Houston executives step out to begin a 24-hour "Management Visibility Tour."

3:00 p.m. - Safety meeting for negative pressure test.

3:04 p.m. - Seawater is pumped into boost, choke and kill lines to displace mud. 1,200 psi is left on kill line.

3:56 p.m. - 424 bbls of 16 ppg spacer followed by 30 bbls of freshwater is pumped into the well, displaced by 352 bbls of seawater. This displaces 3,300' of mud from below the ML to above the BOP. The 16 ppg spacer is placed 12' above BOP.

4:00 p.m. - Tour for executives begins.

4:54 p.m. - Pumps are shutdown, drill pipe has 2,325 psi. Kill line has 1,200 psi. Annular preventer is shut for negative pressure test, isolating the well from downward pressure exerted by the mud and spacer in the riser.

4:56 p.m. - Drill pipe is bled down from 2,325 psi to 1,220 psi to equalize pressure.

4:57 p.m. - Kill line opened and pressure falls to 645 psi. Drill pipe increases to 1,350 psi. Drill pipe is opened to bleed off any pressure. The crew tries to bleed the drill pipe pressure down to 0 psi, 24 bbls are bled off but they cannot get it below 266 psi. Kill line is decreased to 0 psi. Kill line is closed.

4:59 p.m. - The drill pipe is closed; pressure increases from 266 psi to 1,262 psi in 6 minutes.

5:10 p.m. - Crew notices fluid level in the riser is falling. Crew members use a flashlight to peer down into the riser to see fluid level.

Around this time, the night crew gathers for the upcoming shift change. Both Toolpushers and both Company Men are in the driller's shack. The visiting executives also enter the shack as part of the rig tour, area becomes crowded. The executives notice there may be a problem and suggest two key people on the tour help out. The executives then leave. The crew notices fluid level inside the riser dropping, indicating the spacer was leaking down past the annular preventer, out of the riser and into the well. Transocean Offshore Installation Manager orders the annular preventer closed more tightly to stop the leak. Annular preventer closing pressure is increased from 1,500 psi to 1,900 psi. The crew refills the riser with 50 bbls of mud.

5:27 p.m. - Drill pipe pressure is bled down from 1,262 psi to 0 psi by bleeding off ~20 bbls of fluid to the cement unit.

5:52 p.m. - Kill line is bled to the cementing unit. 3 bbls to 15 bbls of seawater are bled back that spurted and was still flowing when instructed to shut-in the line.

6:00 p.m. - Drill pipe pressure increases from 0 psi to 1,400 psi over 35 minutes.

Day-shift Toolpusher is convinced that something isn't right and doesn't believe the explanations he is hearing. But his shift is up. Transocean crew and BP Company Men meet on the rig floor to discuss the readings. It is suggested that the 1,400 psi pressure on the drill pipe is due to a phenomenon called the "bladder effect" or annular compressibility. Night-shift Toolpusher explains that heavy mud in the riser is exerting pressure on the annular preventer, which in turn transmits pressure below. The Night-shift Toolpusher, a highly respected man on deepwater well control is on his last shift on the Deepwater Horizon because he was promoted to teaching at Transocean's well-control school and scheduled to fly out the next day. According to BP witness accounts, he explained that the pressure buildup after bleeding was not unusual. He told the company men, "This happens all the time." The Driller apparently agreed that he had seen the "bladder effect" before. Night-shift Company Man is concerned about the 1,400 psi and continues talking about it to the crew. He says, "TO had dismissed drill pipe pressure as anything serious, somewhat joked about my concern over drill pipe - they found it humorous that I continued talking about [it] in doghouse."

Day-shift Company Man statements to BP investigators suggest that he was present for the discussion as well and that he too accepted the explanation. He justified his acceptance to the investigators by saying that if the Night-shift Toolpusher had seen this phenomenon so many times before it must be real.

Day-shift Company Man later wrote an email to BP Team Leader explaining how the "bladder effect" could account for the 1,400 psi on the drill pipe: "I believe there is a bladder effect on the mud below an annular preventer as we discussed. As we know the pressure differential was approximately 1400-1500 psi across an 18 ¾" rubber annular preventer, 14.0 SOBM plus 16.0 ppg spacer in the riser, seawater and SOBM below the annular bladder. Due to a bladder effect, pressure can and will build below the annular bladder due to the differential pressure but cannot flow – the bladder prevents flow, but we see differential pressure on the other side of the bladder."

TEST YOUR SKILLS

What is wrong with the "bladder effect" explanation?

There is no such thing, it does not exist.[29]

Even if it did exist, any trapped pressure would disappear after bleed-off. If you calculate a differential pressure of 14 ppg mud and 8.5 ppg sea water at 5,000' you get 1,400 - 1,500 psi, which is equal to the pressure seen on the drill pipe, but it is only a coincidence that it is equal.

6:35 p.m. - Night-shift Company Man orders another negative test to be performed. He insists the negative test be repeated on the 3" kill line. Seawater is pumped into the kill line to confirm full. Kill line is opened and bled less than 1 bbl to mini trip tank. Kill line monitored for 30 minutes with no flow. The test on the kill line thus satisfied the criteria for a successful negative pressure test, no flow or pressure buildup for a sustained period of time; however, pressure on the drill pipe remains at 1,400 psi.

7:00 p.m. - Executives gather in the third floor conference room with the rig's leadership. According to BP Vice President, the Deepwater Horizon is "the best performing rig that we had in our fleet and in the Gulf of Mexico. And I believe it was one of the top performing rigs in all the BP floater fleets from the standpoint of safety and drilling performance." BP Vice President, at his new job just four months, was on board in part to learn what made the rig such a stand-out. They had not had a single lost-time incident in 7 years.

> ***TEST YOUR SKILLS***
>
> **Why might the kill line be able to be bled to 0 psi even though the well had 1,400 psi?**
>
> *The spacer migrated into the 3-1/16" kill line clogging it, or mud got in there, or it was clogged by gas hydrates.*[30]

7:55 p.m. - BP Company Men, in consultation with the crew, conclude that no flow from the open kill line yields a successful negative-pressure test, confirming well integrity. Day-shift Company Man heads off duty.

8:00 p.m. - Crew opens the annular preventer and begins displacing mud and spacer from the riser. The Driller sits down in his driller's chair to monitor the well for kicks.

"A reader lives a thousand lives before he dies...
The man who never reads lives only one."
George R.R. Martin, Novelist[31]

How to Monitor for Kicks:

A. **Track volume of fluid in the active pits.** An increase in volume indicates something is flowing into the well.[32]

B. **Track volume & flow rate from the well.** It should equal volume and flow rate pumped into the well. Also track fill-up on trips.[33]

C. **Track pressure.** If drill pipe pressure decreases while pump rate is constant, it could indicate hydrocarbons have entered wellbore, reducing hydrostatic pressure. If drill pipe pressure increases while pump rate is constant, it could indicate heavier mud is being pushed up from below. Unexplained pressure changes may not always indicate a kick, but should always be investigated. Shut down the pumps and monitor the well to confirm it's static.[34]

D. **Perform visual flow checks.** When the pumps are shut off, flow out of the well should stop; however, it generally takes a period of time for flow-out to drop to zero. This reflects the time it takes for the pumps to drain and for circulation to come to a stop. This is called residual flow. Each rig has its own residual flow-out signature, a pattern wherein flow-out dissipates and levels off over several minutes. It's important that you identify that signature and monitor flow-out for a sustained period of time afterward to confirm that there is indeed no flow.[35, 36]

> ***TACTICAL TIP***
>
> If you have any doubt, there is no doubt.
>
> *Shut the well in and monitor the pressure.*

E. **Monitor gas content.** An increase in gas content of fluid returns can indicate an increase in pore pressure, penetration of a hydrocarbon-bearing zone, or a change in wellbore dynamics allowing more effective cuttings removal. However, unexplained increases in gas content can also indicate that a kick is occurring or wellbore conditions are becoming conducive for a kick.[37]

8:34 p.m. - The crew does three things simultaneously:

1) Directs returns away from active pits into reserve pit.
2) Empties the sand traps into the active pits.
3) Begins filling the trip tank .

Each of these actions further complicated pit monitoring for well control purposes. In order to know the volume coming out of the well, the crew had to perform calculations, taking into account that returns were going to two different places: the reserve pit and the trip tank.

8:49 p.m. - The crew again reroutes returns, this time from one reserve pit to another.

9:01 p.m. - Drill pipe pressure begins to slowly increase from 1,250 psi to 1,350 psi, with pump rate constant. The magnitude of the increase would have appeared subtle on the screen, showing only trend lines, but it likely would not have been subtle on the numerical displays.

> ***TACTICAL TIP***
>
> During critical operations, zoom into your digital monitoring charts so you can see small changes in trends. Also, monitor numerical displays closely to identify and address any issues before they become train wrecks.

9:08 p.m. - Pumps are shut down to perform sheen test on returning spacer before dumping overboard. They close a valve on the flow line that had been carrying fluids from the well to the pit system. Mud engineer samples and tests the fluid. Night-shift Company Man waits for confirmation that there is no oily sheen on the returning spacer. Mudlogger performs visual flow check to ensure the well is not flowing while the pumps were off. According to the Mudlogger, there is no flow.

9:10 p.m. - Mud Engineer takes a sample of the returning fluid from the shaker house to the mud lab to run the test. He then returns to the shaker house, weighs the sample, and speaks with another mud engineer about the results. The Company Man gets the results and signs off on the test. With the pumps off, drill-pipe pressure goes up from 1,017 psi to 1,263 psi over the course of 5.5 minutes.

9:14 p.m. - Pumps are turned back on.

9:15 p.m. - Crew begins discharging the spacer overboard.

9:18 p.m. - A pressure-relief valve on one of the pumps blows. The Driller organizes a group of crewmembers to go to the pump room to fix the valve.

9:20 p.m. - Sr. Toolpusher (STO) calls the rig floor and asks Night-shift Toolpusher (TO) about the negative-pressure test.

TO: "It went good. . . . We bled it off. We watched it for 30 minutes and we had no flow."
STO: "What about your displacement? How's it going?"
TO: "It's going fine. . . . It won't be much longer and we ought to have our spacer back."
STO: "Do you need any help from me?"
TO: "No, man. I've got this. Go to bed. I've got it."
STO: "Okay."

9:27 p.m. - Drill crew opens the kill line valve. Kill line pressure increases to 800 psi. Drill pipe pressure is at 2,500 psi. The Driller notices the odd and unexpected pressure difference between the drill pipe and the kill line.

SURVIVAL SKILLS: LIFE, DEATH, AND MONEY

When things are unclear in this business, various reasons and explanations for the uncertainty will be presented. Often they will come from different people on location, offering an explanation to make sense of the situation. It seems everyone has a strong opinion when dealing with ambiguity, especially when working on a problem thousands of feet beneath the surface of the Earth. Sometimes it's hard to know who or what is correct.

Various explanations for the uncertainty will also come from your own mind. In my experience it is not uncommon to agree with the explanation that makes your life easiest. Nobody wants to have problems or troubled wells. And it is far too easy to dismiss problems by ignoring them, hoping they will go away, justifying warning signs with theories, or explain them away in your mind. Under the stress of uncertainty, your mind will play tricks on you.

Sometimes you can talk yourself into agreeing with the explanation that you want to be true, regardless of what the data shows. If you have worked in this business for any significant length of time, this has happened to you, including the author of this book. Additionally, when the aspect of cost comes into the picture, there is a certain type of pressure that is hard to put into words, especially if it is your own personal finances being vaporized due to a delay or problem on location.

Clear your mind. Consider calling someone off location to discuss the issue. Always have a contact list of key people that you can call 24/7 to discuss things. Getting a second opinion from someone who is not involved with the job or under the stress of dealing with the uncertainty can be a life saver, and can also save your company a massive amount of money; like $60 billion. . .

"There is no difference between 'just a small flow' and a 'full flowing' well
because both can very quickly
turn into a big blowout."
Neal Adams, Well Control Expert[38]

9:30 p.m. - Crew shut off the pumps to investigate. Drill pipe pressure initially decreases after the pumps are turned off.

9:32 p.m. - With the pumps off, drill pipe pressure increases from 1,210 psi to 1,766 psi over a 5 minute period. Meanwhile, the pressure on the kill line remains significantly lower.

9:36 p.m. - The Driller orders a Floor Hand to bleed off drill pipe pressure, in an apparent attempt to eliminate the difference.

9:37 p.m. - Pressure drops off to ~500 psi; however, once he closes the drill pipe valve, pressure shoots up to 1,450 psi.

TEST YOUR SKILLS: PRESSURE

What are potential reasons for the pressure increase?

Communication with the reservoir.

Hydrocarbons have entered the wellbore due to:

1) Flow up the annulus and through the seal assembly. [39]
   - *Debris obstructed the seal during the setting process*
   - *Seal failed to expand and set properly*
   - *Seal dislodged after it was set*
   - *Pressure & forces lifted casing hanger up out of place*

2) Flow into the long string casing. [40]
   - *Shoe cement failed and floats are not closed or not holding*
   - *Failure of 9-7/8" x 7" crossover*
   - *Casing connection leak*
   - *Hole in the casing*

9:38 p.m. - Despite evidence of a kick, a visual flow check is not performed and the well is not shut-in.

9:39 p.m. - Drill-pipe pressure changes direction and slowly decreases from 1,400 psi to 338 psi. In retrospect, this is a very bad sign. It likely meant that lighter-weight hydrocarbons were now pushing heavy drilling mud out of the way, up the casing, past the drill pipe.

9:40 p.m. - Drilling mud begins spewing from the rotary onto the rig floor. Toolpusher and assistant driller return to the rig floor. They route the flow coming from the riser through the diverter system, sending it into the mud-gas separator.

9:41 p.m. - Toolpusher activates the upper annular preventer. (Note: upper annular generally takes 26 seconds to close on Deepwater Horizon and may have been closed on a tool joint which may have prevented the annular from forming a good seal around the pipe. Forensic analysis of recovered equipment indicates a tool joint in the upper annular region contained significant erosion.)

9:45 p.m. – Assistant Driller calls Sr. Toolpusher to tell him that the well is blowing out, mud is going to the crown.

9:47 p.m. - Rig crew activates upper pipe ram variable bore ram (VBR) and middle pipe ram VBR. VBRs generally take 16 seconds to close on Deepwater Horizon. Flow rates at this point may have been too high for the annular preventer or VBRs to fully seal. By the time the crew acted, gas was already above the BOP, rocketing up the riser, and expanding rapidly. However, there is evidence the VBRs successfully sealed the annulus. Hydrocarbons were probably also travelling up the drill string.

9:48 p.m. - Gas hissing noise, first gas alarms sound. Roaring noise and vibration. Drill pipe pressure rapidly increases from

1,200 psi to 5,730 psi (the significant drill pipe pressure increase is evidence the VBRs sealed the annulus)

9:49 p.m. - Rig loses power. 1st explosion occurs.

9:52 p.m. - "Mayday, Mayday, Mayday, this is the Deepwater Horizon. We are on fire."

9:56 p.m. - Attempt to activate emergency disconnect sequence (EDS). Lower marine riser does not unlatch.

The EDS should activate the BSR and sever the drill pipe, seal the well, and disconnect the rig from the BOP. It is possible that the first explosion already damaged the cables to the BOP, preventing the disconnect sequence from starting. There is evidence that there were no hydraulics. Communication to the BOP was lost either in the explosion and/or fire. However, the BOP's automatic mode function (AMF) "deadman" system should trigger the BSR after the power, communication, and hydraulics connections between the rig and the BOP were cut. In dispute are a low battery charge in the blue control pod and incorrectly wired solenoid valve in the yellow control pod which may have impacted the AMF.

Analysis from the recovered blowout preventer suggests the BSR did close on April 20th or on April 22nd when the BSR was activated by an ROV. The well continued to flow possibly because high-pressure hydrocarbons flowed around crumpled drill pipe in the BSR which did not get a clean cut. The BSR may have been unable to center the drill pipe due to forces from formation flowing pressures which buckled the drill pipe or forces from above as a result of falling equipment at the surface.

Either situation may have put the drill pipe in compression. This makes the drill pipe harder to cut because the drill pipe may buckle as soon as the BSR attempts to cut it, and the drill pipe will not move out of the way after it is cut which could prevent the BSRs from sealing. There is evidence that the drill pipe was pushed under great force to the side of the BOP, outside the BSRs ability to center the drill pipe, fully shear and seal. DNV was chosen to perform the forensic analysis of the recovered BOP. They concluded that the BSR could not shear the drill pipe and seal the well because the drill pipe was not centered. Additionally, there is analysis that the design of the BSR could be faulty and cannot shear and seal 5 ½" drill pipe in compression. Furthermore, the blades on the BSR do not fully span the BOP. There are gaps on the sides of the blades. The BSR design is the straight blade and V-blade combination which may be inferior to the double-V blade design.

9:57 p.m. - On the Bankston supply vessel, the Captain is on the bridge, updating his log. He steps out and sees "mud falling on the back half of my boat, kind of like a black rain." He calls the Deepwater Horizon bridge, "I'm getting mud on me." Deepwater Horizon calls back and says to move back 500 meters. The Bankston's relief chief and two others launch the fast rescue craft to pick up people that jumped from the rig into the sea.

10:00 p.m. - **11:22 p.m.** - Escape of 115 personnel to the Bankston

10:25 p.m. - Traveling blocks fall

11:30 p.m. - Managers take a final muster. 11 men are missing.

April 21, 2010

1:30 a.m. - Rig lists and rotates in the wake of secondary explosions. Work boats, spraying water on the rig in response to the mayday call, move back.

2:50 a.m. - Deepwater Horizon had spun 180 degrees, with its dynamic positioners dead, the rig moved 1,600 feet from the well.

3:15 a.m. - U.S. Coast Guard cutter Pompano arrives. Search helicopters survey one sector after another.

8:13 a.m. – Bankston is given permission to return to Louisiana.

- 115 survivors
- 11 fatalities
- 16 are seriously injured and medevac'd to hospitals
- 99 are transported to land by the Bankston

April 22, 2010

1:27 a.m. - Bankston arrives at Port Fourchon. Per federal regulations, each person is drug-tested.

"I know it is hard to understand, but sometimes painful things like this happen. It's all part of the process of exploration and discovery. It's all part of taking a chance and expanding man's horizons. The future doesn't belong to the fainthearted; it belongs to the brave."

Ronald Reagan, U.S. President[41]

Methods Of Escape

115 out of 126 people on Deepwater Horizon were able to escape the disaster. The majority of people escaped by lifeboat, which was attributed to key crew members who delayed launching the lifeboats until they boarded as many people as possible.[42] Many acts of heroism occurred during the escape and contributed to the high survival rate.

- 100 people escaped on two lifeboats
- 8 people escaped by jumping into the sea
- 7 people escaped on a raft

Takeaways

There are many lessons from the Deepwater Horizon disaster. If you remember nothing else, remember this:

NEVER: Think these things cannot happen to you.

If you read through the accident timeline and found yourself thinking, "I would never do that. I would catch that issue. I am smarter. They were so stupid," you are setting yourself up to have a train wreck. What happened at Macondo can happen to anyone. You have to think like that to ensure you are always on your toes and prepared for anything.

It is not uncommon for operating teams to let their guard down after the production casing cement job. It is not uncommon in this business to think the risk of a well control situation is eliminated after the cement has been pumped, especially if you believe you had full returns during the job.

"Cement is not a barrier until it is successfully tested by the operator, either with a CBL and/or a negative pressure test... Obviously you have to test it before you can rely upon it to be a barrier."

Don Godwin, O&G Trial Lawyer[43]

2. NEVER: Hesitate to Shut Down an Unsafe Job

People do unsafe things in this business and get lucky; they get away with it. That does not mean they will continue to get away with it. If you see an unsafe operation, you are obligated to address it. You own it. Take responsibility for protecting yourself and your Company by stopping unsafe activity.

Individual work that is unsafe is fairly straightforward to shut down. People generally hesitate when an entire operation needs to be shut down for safety reasons. My experience regarding this matter is that people do not want to shut it down - it's too much trouble. What if they are wrong? Nobody wants to be "that" guy known to shut jobs down and perceived as a trouble-maker or impediment to getting work done. However, in most cases, you don't have to shut the job down and never complete it. In many situations, all it takes is to shut the job down for a few minutes until the issues are addressed. Nevertheless, sometimes addressing safety issues can take a while, causing a delay and financial impact. Don't worry about the cost. Shut the job down, ensure everyone is safe, then worry about the money. An accident is always more costly.

SURVIVAL SKILLS

One of the greatest responsibilities you are given in the oil and gas industry is to stop unsafe operations. Whether you have 1 day of experience or 30 years, you are obligated to take action.

If someone shuts a job down, never override that person. Once a job is shut down for safety, everything must be addressed before continuing operations. The person that shut the job down should be respected and consulted before work is continued. If you exercise Stop Work Authority to shut a job down and someone tries to override you or dismiss your concern, do not back down. Protect yourself and your Company by challenging them.

"Those who have the privilege to know, have the duty to act."
Albert Einstein, Physicist[44]

Let's review three oilfield stories, where significant adversity is experienced when Stop Work Authority is utilized on location. It takes great courage to shut a job down.

Pump Hand Shuts Job Down Saving 54 Lives

On a remote international location, a natural gas shale exploration well was drilled, cased, and cemented. The drilling rig moved off location, and completion operations were mobilized to the site, including 54 people involved with the operation. The first step was to perforate the lower interval and attempt to flow the well naturally. Since this was a shale exploration well, there was uncertainty regarding whether it would flow without a stimulation treatment. The interval was perforated around 3 p.m., and flow testing commenced. Around 11 p.m., the flowback crew noticed the concentration of H_2S was increasing from around 10 ppm to over 100 ppm within a time span of two hours. The flowback supervisor woke up the Company Man on location and told him about the H_2S.

Flowback Supervisor: *"We have 100 ppm H_2S on the well."*

Company Man: *"Are you sure?"*

Flowback Supervisor: *"Yes, double checked with different detectors."*

Company Man: *"Let's go take a look."*

Company Man and Flowback Supervisor walk over to the flowback trailer, confirming the H_2S concentration.

Company Man: *"Let me call the boss, hold on."*

Company Man phones Engineer and Asset Manager at the base.

Company Man: *"We are flowing back H_2S, 100 ppm."*

Oil Company Engineer: *"Are you sure? There is no H_2S in this country."*

Company Man: *"I think we just discovered the first well."*

Oil Company Engineer: *"It should cleanup. I bet it's just sulfate reducing bacteria that built up and needs to be flushed out. I have seen this before. The concentration should drop. Keep flowing it back."*

Company Man: *"Okay, boss."*

The next morning, the frac crew starts rigging up to frac the well, when one of the pump hands overhears a conversation between the Frac Engineer and Company Man talking about the H_2S. The Pump Hand asks the Frac Engineer about the situation.

Pump Hand: *"Are we flowing H_2S out here?"*

Frac Engineer: *"Yes, that is what it looks like."*

Pump Hand: *"What concentration?"*

Frac Engineer: *"It was 100 ppm, but it has been increasing and is currently over 200 ppm. They think it will cleanup though."*

Pump Hand: *"How do they know that? This isn't good. We're not prepared to deal with H_2S."*

Pump Hand inspects the wellhead and talks to the wellhead and stack crew. He realizes the equipment is not rated for sour service. The Pump Hand then talks to the Frac Service Leader on location.

Pump Hand: *"We have over 200 ppm flowing back right now."*

Service Leader: *"Yeah, I heard about it."*

Pump Hand: *"We are not prepared for this. We don't have enough H_2S detectors, we have no SCBAs except one unit that looks like crap from the 1970's and the wellhead is not sour service rated."*

Service Leader: *"Who said the wellhead is not sour service rated?"*

Pump Hand: *"The wellhead supervisor."*

Service Leader: *"It's only 200 ppm."*

Pump Hand: *"I'm shutting the well in and shutting this job down. We need to get the right equipment on location to deal with H_2S."*

Service Leader: *"Shutting the job down? Don't you think the Company Man and Engineers would shut the well in if they thought it was a problem? We are going to frac the well anyway. They will close the well before the frac in a few hours."*

Pump Hand: *"What if something happens during the rig up, frac job, or flowback after the frac job, then what?"*

Service Leader: *"You are going to make yourself look like a fool. Probably get fired, not by me, but back at the office. They don't like this sort of thing, shutting jobs down. Are you crazy?"*

Pump Hand: *"I don't care. I'm doing it. I am telling you right now that I'm shutting this job down. Shut the well in right now!"*

The Pump Hand walks over to the Frac Engineer and Company Man and tells them to shut the well in.

Company Man: *"Why should I shut the well? And who are you?"*

Pump Hand: *"I'm Mike, a Service Operator with Honor Frac Services, and we are not prepared to deal with H_2S. We need to get the proper equipment. Plus, your wellhead is not sour service rated."*

Company Man: *"The mangers at the Oil Company are saying this H_2S is going to clean up and should not be an issue, so calm down. Go talk to your superiors. Who is this guy? Get this guy out of here."*

The Frac Engineer, realizing that someone on location just exercised their authority to Stop Work, tells the Flowback Supervisor to shut the well in. He then talks with the Service Leader.

"All personnel are empowered and obligated to 'stop a job' that puts people or the environment at risk."

Dave Dunlap, CEO Superior Energy Services[45]

Frac Engineer: *"Mike is shutting the job down, initiating Stop Work."*

Service Leader: *"Yeah, I know, he looks like a fool."*

Frac Engineer: *"Does not matter, we need to respect it and shut the well in, call the Country Manager to notify him of the situation. Plus, he is right, we are technically not supposed to be dealing with H_2S without the proper equipment. It's against company procedure."*

Service Leader: *"I know. It's going to create a shit storm."*

The well is shut-in even though the Company Man is yelling to not shut the well in. The Service Leader calls the Country Manager informing him of the situation. After much discussion, arguing, and yelling, mostly by the Oil Company leadership who are losing $250,000 per day every day the operation is on standby, the decision is to have everyone leave location except for a small crew, including the Company Man and a few flowback hands.

Fifty four people leave location and travel back to the operations base. The plan is to return to location when the proper equipment is mobilized in-country to handle high concentrations of H_2S. Mike the Pump Hand ends up being made fun of by several people on the frac crew for shutting the job down. They call him a wimp and suggest he should get out of the oil business. On the long trip back to the base, after being ridiculed by the frac crew for hours, Mike starts to doubt whether he made the right decision to shut the job down for safety.

Once everyone is off location, the Oil Company ignores the concern expressed by Honor Frac Services personnel and decides to open the well and continue flowing it back. Over the course of several days the concentration of H_2S continues to increase until it is over 1500 ppm. Above 1000 ppm is considered "nearly instant death."

"You have the right to ask questions, the obligation to stop an unsafe act by another employee or contractor, and the ability to refuse to perform a task that is either unsafe or for which you have not been properly trained."

Nathan Houston, CEO U.S. Well Services[46]

Hydrogen Sulfide (H_2S)[47]

Concentration (ppm)	Symptoms/Effects
0.00011-0.00033	Typical background concentrations
0.01-1.5	Odor threshold (when rotten egg smell is first noticeable to some). Odor becomes more offensive at 3-5 ppm. Above 30 ppm, odor described as sweet or sickeningly sweet.
2-5	Prolonged exposure may cause nausea, tearing of the eyes, headaches or loss of sleep. Airway problems (bronchial constriction) in some asthma patients.
20	Possible fatigue, loss of appetite, headache, irritability, poor memory, dizziness.
50-100	Slight conjunctivitis ("gas eye") and respiratory tract irritation after 1 hour. May cause digestive upset and loss of appetite.
100	Coughing, eye irritation, loss of smell after 2-15 minutes (olfactory fatigue). Altered breathing, drowsiness after 15-30 minutes. Throat irritation after 1 hour. Gradual increase in severity of symptoms over several hours. Death may occur after 48 hours.
100-150	Loss of smell (olfactory fatigue or paralysis).
200-300	Marked conjunctivitis and respiratory tract irritation after 1 hour. Pulmonary edema may occur from prolonged exposure.
500-700	Staggering, collapse in 5 minutes. Serious damage to the eyes in 30 minutes. Death after 30-60 minutes.
700-1000	Rapid unconsciousness, "knockdown" or immediate collapse within 1 to 2 breaths, breathing stops, death within minutes.
1000-2000	Nearly instant death

With the well flowing 1500 ppm for several days, the Oil Company starts to question the analysis suggesting the H_2S would cleanup. They begin to accept that the natural gas exploration well is a sour gas discovery well. The decision is made to shut the well in and prepare for the stimulation treatment once the H_2S equipment arrives in country. It was being flown in from the United States.

There is uncertainty as to what happened next. It is believed that once the well was shut-in, a leak developed at the wellhead below the lower master valve. The Cook called the base when the Company Man and flowback hands did not return for dinner one evening. A fire was noticed by the Staff Manager the next day. Well control was mobilized to location. It is believed the Company Man and flowback hands tried to escape, as they were all found face down and dead not far from the wellhead. An autopsy revealed everyone on location died from hydrogen sulfide poisoning.

Case Analysis

Mike the Pump Hand exercised Stop Work Authority and was credited with saving the lives of the 54 people who left location once it was determined that the proper equipment was not available to perform the job safely. Mike the Pump Hand is considered a hero by the company, his colleagues, and their families for his actions. After years of impressive work service, he was promoted to service leader, then country manager, where he currently works and is responsible for the safety and success of a team with over 200 employees.

"A man must know his destiny... if he does not recognize it, then he is lost.
By this I mean, once, twice, or at the very most, three times,
fate will reach out and tap a man on the shoulder...
if he has the imagination, he will turn around
and fate will point out to him what fork
in the road he should take,
if he has the guts,
he will take it."
General Patton[48]

Wireline Operator Saves Life of Family Man

The operation was to run several wireline logs to evaluate the reservoir. The Wireline Operator noticed that the Crane Operator was lifting the lubricator and tools over several of the ground crew personnel. On one occasion, it was directly over the head of the Wireline Engineer, who was also the boss on location. However, nothing was said and the operation continued. Additionally, random people on location were walking under the overhead load.

After wireline operations were finished, the Crane Operator began another project for the Toolpusher and Driller. They were moving several pieces of equipment for the rig mob to the next location. The Wireline Operator, seeing that the Crane Operator continued to move overhead loads over people, walked to the Crane Operator and asked him to put the load down and stop working for a few minutes to talk. The Toolpusher noticed the Crane Operator had put the load down, so he walked over to the crane to find out what the problem was.

Toolpusher: *"What's the problem?"*

Wireline Operator: *"You're operating the crane unsafely, so I shut the job down. Only need a few minutes to discuss this issue . . ."*

Toolpusher: *"It's not your business. You are done with wireline operations, now get off location."*

Wireline Operator: *"Listen, I'm concerned for your safety and the safety of your crew."*

Toolpusher: *"We are fine. Mind your business."*

Wireline Operator: *"Did you notice they are moving equipment directly over your head? If something happens, you're gonna get crushed."*

Toolpusher: *"How long have you been in the oilfield?"*

Wireline Operator: *"Two years. Do you have a problem with that?"*

Toolpusher: *"I have been doing this for over 30 years."*

Wireline Operator: *"So what? Doesn't mean I'm wrong."*

Toolpusher: *"If it's going to fall, it's going to fall."*

Wireline Operator: *"But you don't have to be under it."*

Toolpusher: *"If I am going to die in God's oil patch, then it's going to happen regardless of what I do. When your time is up, it's up. There ain't nothing you can do about it."*

Wireline Operator: *"Do yourself a favor, old man, and don't stand under overhead loads. And to you Crane Man, no more moving loads over the top of people."*

Crane Man: *"Whatever. Okay. Got it."*

Wireline Operator: *"Thank you. Continue your operation."*

Toolpusher: *"We will continue our operation. Mind your damn business! And get the heck off my location, worm."*

The Wireline Operator walked away, preparing to leave location. Crane operations continued. After 15 minutes, they were moving a large piece of equipment when the crane hook failed, dropping the load right next to the Toolpusher. After the dust settled, the Toolpusher walked over to the Wireline Operator and thanked him for saving his life.

Toolpusher: *"You saved my life. Sorry I was such a jerk earlier. I'm a mean old bastard. That's what years away from your family will do. I have 6 kids, 3 are under the age of 10. Thank you from the bottom of my heart. Thank you for saving my life."*

Wireline Operator: *"Thank you for listening to me."*

Toolpusher: *"I didn't listen, it was the Crane Man who listened. I'm a fool. I don't know what my family would do without me. God sent you here to this location to save me."*

To this day, the two are friends. Whether Stop Work Authority saves the life of one person or 1,000 people, it is the same to the families of the folks in the field. At the end of the day, family is the most important thing. When you think about it, that's who we really work for. If you are in a situation where you're hesitant to shut a job down, think about what people's family would want you to do.

Stop Work Authority Overruled By Frac Manager

The operation for the day was a vertical well single stage frac job. Before the frac, Loose Canon Corporation wanted to pump an acid ball out. The acid pumper, Ivan Ironjaw, was surging balls off of perfs when a long radius chicksan swivel failed, probably due to hydraulic shock from the pressure surge, also known as the water hammer effect. When the swivel joint failed, the iron swung around and hit Mr. Ironjaw directly in the face. Always Pumping Services frac crew witnessed the accident. Ivan was rushed to the hospital. This shook up the frac crew and everyone on location, except the Company Man, Brian Budgets, who was also the majority owner of the well and a financial analyst by trade.

The Blender Tender, Mr. Braveheart, came down from the blender and walked over to the customer, Mr. Budgets, the Head Frac Leader Edward Dangerhands, and a few other people in leadership positions who were congregated, discussing the failed iron situation. Mr. Braveheart told them he was shutting the job down and that they would not frac today. Edward immediately got in Mr. Braveheart's face and challenged him as to why he was shutting the job down.

"Every person has the right and responsibility to stop any task that is believed to be unsafe or could lead to environmental impact."

Doug Lawler, CEO Chesapeake[49]

Edward said, *"We work for 'Always Pumping' and our slogan is that we always pump, all day every day!"* Braveheart walked off, not wanting to deal with him. Edward assured Mr. Budgets that they would frac today as Mr. Budgets reminded everyone that, *"Every hour the operation is on standby I am losing $10,000. I will be over budget. This wasn't in the plan that I have in my budget book. By God, you must pump this frac today!"* Edward called the New Mexico camp office to inform Always Pumping Services Vice President, Mr. Henry Hatchet, of the situation. Mr. Hatchet told Edward to see if there was a way to continue operations safely. For the next several hours Edward tried everything to get Mr. Braveheart and the frac crew, who all agreed on exercising Stop Work Authority, to pump the frac. Mr. Braveheart and the frac crew insisted that they would not frac until all of the iron was inspected.

Heated discussions turned to yelling and disrespect as the situation escalated. Mr. Budgets got involved and insulted the entire frac crew saying, *"You are nothing to me, just hands that will be replaced tomorrow. I have all of your names, and none of you will be allowed on my locations. I will tell everyone in Southeast New Mexico and West Texas that you are no good worms, all of you!"* Then Henry Hatchet, Pumping Services VP, showed up on location. He told the frac crew that the iron was good and that they would pump the job or seek employment elsewhere. The crew questioned how he determined the iron was good. Henry told them the acid pumper iron and frac iron were not the same pressure rating and two different sets of iron. The crew already knew this, but they still did not feel comfortable after witnessing the accident.

After listening to the Lead Treater, Vice President, and Company Man for hours, screaming to pump the job or get replaced and essentially fired, the crew pumped the job despite not wanting to do it. Pumping commenced and the job was pumped without incident.

"The ultimate measure of a man is not where he stands in moments of comfort and convenience, but where he stands at times of challenge and controversy."

Martin Luther King Jr., Civil Rights Leader[50]

News of Stop Work Authority being overruled on location by area management made its way through the grapevine to the Houston office. Always Pumping Services EVP of Operations commissioned an investigation into the matter. After a few weeks, the investigation was concluded. The frac iron was inspected and found to have had several issues. The Frac Leader Mr. Edward Dangerhands and VP Mr. Henry Hatchet were terminated for overriding Stop Work Authority. Loose Canon Corporation was banned from Always Pumping Services accepted client list.

Case Analysis

As this case illustrates, never overrule Stop Work Authority or force someone to perform an operation if they do not feel comfortable or safe. As a result of this incident, the crew lost faith that the company cared about their safety. The Stop Work Authority program lost all credibility and was considered a joke among the employees. A number of people quit as a result.

The frac crew was shook up after the near fatal accident on location. They were not in the right state of mind to pump the job. The investigation determined a 24-hour safety stand-down should have occurred after the failed acid pumper iron accident to assess the situation. The Blender Tender who initiated Stop Work Authority and stood up to the Frac Leader and Vice President was rewarded by the EVP of Operations for his actions. He became a legend and hero to all those who knew what happened.

"Will you be the rock that redirects the course of the river?"

Claire Nuer, Conflict Resolution Expert[51]

3. ALWAYS: Protect the Environment

Safety is always *NUMERO UNO*. Environment is a close second. Some folks argue that environment is just as important as safety, and that looks great on paper. However, in the real world, safety will always be more important than protecting the environment because one cannot protect the environment if one is dead or about to die. Get your safety right, then focus on the environment.

In most situations, you will not have to choose between safety and the environment. They go hand-in-hand, as safety and the environment are closely connected. By protecting the environment, you are almost always creating a safer operation. Therefore, always take efforts to protect the environment just as seriously as safety-related issues while maintaining your priority pledge. In fact, in today's world, environmental accidents often get a lot more attention from the media compared to fatalities. One area that has recently received a significant amount of attention is salt water disposal (SWD) operations.

Justice Department Indicts SWD Operator

In 2015, a man (Man #1) was indicted in a federal court on 13 felony charges surrounding the operation of a saltwater disposal (SWD) well in North Dakota. According to the indictment, Man #1 "conspired with others . . . in a number of coordinated and illegal acts, including injecting saltwater into the well without first having the state of North Dakota witness a test of the well's integrity and continuing to inject saltwater after failing a pressure test."[52] Man #1 "is also charged under the Safe Drinking Water Act with injecting fluids down the 'annulus' or 'backside' of the well in violation of the well's permit which required that fluids be injected through the tubing."[53]

"Have the courage to say no. Have the courage to face the truth. Do the right thing because it is right. These are the magic keys to living your life with integrity."
W. Clement Stone, Businessman[54]

Additionally, Man #1 is "charged with telling Man #2 to move a device called a 'packer' up the wellbore in violation of the well's permit, without first getting approval from the state."[55] Man #2 pleaded guilty in a federal court in 2014 to 11 felony charges stemming from the operation of the SWD well.[56]

SWD Operator Indictment Key Points

"Environmental Protection Agency (EPA) has defined five classes of injection wells, with Class II wells receiving certain injected fluids related to oil and natural gas production. Class II wells receive brine and other wastes commonly referred to as 'saltwater.'

Under the Safe Drinking Water Act (SDWA), states are the primary enforcers of the Underground Injection Control (UIC) program. Once a state program meets minimum federal standards, it may secure primary enforcement authority for the regulation of underground water sources if the EPA approves the state's UIC program. North Dakota has an EPA-approved UIC program.

When a state obtains primary enforcement authority, the federal government retains enforcement authority, including the right to initiate criminal charges for violations of the SDWA. **It is a crime under the SDWA for a person to willfully violate any requirement of an applicable UIC program.**

- North Dakota regulations prohibit underground injection into Class II wells without a permit. . . . Such a permit 'may contain such terms and conditions as the commission deems necessary.'
- A 'Mechanical Integrity Test' (MIT) includes a positive pressure test that requires a well to hold pressure without leaking and ensures ground water is protected. . . . Operators are required to demonstrate continual mechanical integrity.
- A device called a 'packer' is used to isolate the injection zone from the space between the tubing and injection casing above

the packer, called the 'annulus.' . . . A packer serves as an extra layer of protection for ground water by providing a seal between the outside of the tubing and the inside of the casing to prevent the movement of fluids. The hydraulic seal provided by a properly set packer and the cement above it prevents injected material from migrating up the wellbore into a fresh ground water zone and causing pollution. North Dakota regulations require that wells be 'equipped with tubing and packer set at a depth approved by the director.'

- The amount of injection for each well must be reported monthly.
- Annular injection of fluid is prohibited."[57]

Overt Acts Included in the Indictment:

- "Converted well to SWD; didn't notify the North Dakota Industrial Commission (NDIC) within 30 days.
- Commenced injection but didn't immediately notify NDIC.
- Did not file a monthly report of the amount and sources of the fluid injected.
- Injected saltwater down the annulus.
- Injected saltwater without an NDIC Field Inspector witness a satisfactory MIT on the tubing-casing annulus.
- Conducted a pressure test which failed.
- Injected 8,000 bbls in violation of an NDIC verbal shut-in order that the well could not be used until an Inspector witnessed an MIT.
- Falsely stated in an email, 'we did not take any water at all during the time we were waiting for the mechanical test' and 'no pumping done for commercial gain during this time.' Admitted only turning the pumps on 'briefly every 12 hours so the pipes don't freeze'.

- Reset the packer, moving the packer up hole more than 100 feet from the top perforation.
- Ordered crew to move all of the joints that had been taken out of the well to behind a shack to conceal them from the NDIC.
- Sent an email to the NDIC with a wellbore schematic stating the packer is at 5,546 feet; however, the packer was not at 5,546 feet.
- Falsely stating, 'the rig that was on site merely released the packer and re-set the packer because it slightly leaked at 900 psi, after it was re-set it held great' "[58]

Case Analysis

As this case illustrates, if you do not protect the environment and take environmental regulations seriously, you are effectively playing with your personal freedom. Sooner or later you will end up in a legal situation. It is just not worth it. When working on SWD related issues, consider hiring a professional and/or having several experienced engineers, geologists, and geophysicists review and validate all activities surrounding SWD construction and operations. Additionally, consider frequently auditing every aspect of your SWD system — from well activities to record-keeping — to ensure operations are continually in compliance.

"Earth provides enough to satisfy every man's needs, but not every man's greed."

Mahatma Gandhi, Anti-War Activist[59]

4. CREATE: Value Everyday

Price is what you pay for something, value is what you get in return. While attending business school at Columbia University, I had a professor who would repeat "price is what you pay, value is what you get" over and over in each class. It's a quote made famous by Warren Buffet, the billionaire investor, with a net worth of $71 billion and one of the world's most powerful people, according to Forbes 2016 ranking. Buffet also learned "price is what you pay, value is what you get" from a Columbia professor, Ben Graham, who is considered the father of value investing. The concept of investing in something when you get more than what you pay for helped Buffet make billions and it translates directly into oilfield survival.

Whether you work for an oil and gas operator, a service company, or as an independent consultant, never forget that your employer has made an investment in you because they believe that they are getting more value than holding onto their money. It is not a given that your job exists. Many people forget this simple truth. Your job exists because the position, whether it's filled by you or someone else, creates more value than the total cost of employment. Let's consider an example in the oil patch.

Low Value Drilling Consultant Gets Fired

Joe Bob Lunchbucket, drilling consultant in the Bakken with 30 years of oilfield experience, was hired to oversee the spud to rig release of 4 horizontal Bakken wells at a flat rate of $1,950 per day. The oil company that hired Joe in essence made a $117,000 investment in Joe.

Mr. Lunchbucket managed location operations, coordinated with engineers and vendors to get the wells drilled without any safety incidents or major operational issues.

"My advice is to work with dedication, distinction, and expertise at a high level of effort on a team basis every day in every way."

Aubrey McClendon, Shale Pioneer[60]

On the surface, it looks like Joe delivered significant value for the $117,000 investment — the wells were drilled on budget with no issues, right?

Joe Lunchbucket: *"I did my job, damn it! The wells are on budget and no incidents. What more do you, sons-of-bitches, want?!"*

Drilling Manager: *"Your daily drilling reports lack detail and were low quality, you ignored the geologist requests regarding staying within the target interval, several near misses went unreported, and costs were not documented correctly. Field estimated cost is off by 30% from actual invoices."*

Joe Lunchbucket: *"I am not a secretary or accountant. You paid me to deliver 4 wells in 60 days. That is what you got. I did my part."*

Drilling Manager: *"No, Joe, you are living in the past, pre-horizontal shale oilfield, the old oilfield. Times have changed, we demand and expect more from you. Also, what happened with the geologist, you threw him off location?"*

Joe Lunchbucket: *"That rock doctor does not know what he is doing, going to get us stuck out here. I had to do it. I'm not trying to twist off. That geologist is going to get us planted, you understand."*

Drilling Manager: *"Joe, you have to learn to be respectful and work with people on the team. This well may end up below type curve considering the lateral is partially outside the target window. Then I am going to have the geology manager and the reservoir manager all over me. There is a way to drill the lateral within the target window and not get stuck."*

Clearly, the Drilling Manager does not feel like he is getting value from Joe. Doing the bare minimum and disrespecting teammates is not acceptable in today's oilfield.

While drilling those four Bakken horizontal wells, Joe identified and addressed several critical aspects that prevented operational train wrecks when drilling the surface, drilling the curve, and running casing. However, Joe never documented his findings or shared this knowledge with the drilling engineers so that it could be incorporated into future procedures.

One month later, an offset drilling crew experienced a $2.0 million train wreck that could have been prevented with Joe's knowledge that he gained when working the offset pad. It was discovered after the Drilling Manager visited the rig on Joe's days off and learned of the successful techniques employed from the floor hands and junior drilling consultant, whom Joe told not to say anything to the office personnel about what they were doing to be successful. Joe was later questioned about this:

Drilling Manager: *"Joe, you are our most experienced drilling consultant. We respect your knowledge. You have excellent spud to rig release times of less than two weeks. For a Bakken well, that's pretty good."*

Joe Lunchbucket: *"Thanks, DM."*

Drilling Manager: *"I'm sure you heard Greg Goodhand got stuck running casing on the offset pad. They could not get to bottom. Ended up cementing in place 1,000 feet short."*

Joe Lunchbucket: *"Yeah, that's too bad. I should oversee them also. Greg's a good guy but he needs help sometimes."*

Drilling Manager: *"Visited your rig the other day, spoke with the guys on location. They mentioned a few of your techniques that consistently work in this area of the Bakken. I questioned them as to why none of this was documented or shared with the engineers. Why is that, Joe?"*

Joe Lunchbucket: *"Ah. Well, it works for me, might not work for others. Those are my methods, part of what I bring to the table, why you pay me higher than the others."*

Drilling Manager: *"$2 million dollars was lost and it could have been prevented."*

Joe Lunchbucket: *"This is job security for me, I mean I can't give away all my secrets. You won't need me then."*

Drilling Manager: *"Hiding information and potential solutions? That's not job security, it's shortsighted, see the bigger picture. We are looking for someone to oversee all 10 of our rigs in North Dakota, your name was mentioned; but hoarding information is not how we do things. If all you do here is the bare minimum, I won't need you now, let alone give you a promotion."*

Operations Analysis

Joe Lunchbucket ended up losing the promotion and ultimately getting fired. Greg Goodhand, an honest man having a solid reputation for consistently delivering more than expected, is currently still employed. When tough personnel decisions are made during industry slowdowns, consistently delivering dependable high value work product will significantly increase the probability that you will survive the cut. After all, this is the Oilfield Survival Guide.

Passing time and doing the bare minimum is not creating value. The bare minimum is just a paycheck. Not only are you destroying value for your employer, but you are also destroying value for yourself. Go above and beyond. Get more involved and engaged. Furthermore, withholding information to benefit yourself at the expense of your employer, as Mr. Lunchbucket did, is not respectable. Once your reputation in the oilfield is damaged, it's hard to fix it.

"If anybody perceives you to be slick or dishonest or crooked, they're not gonna mess with you."
Clayton Williams, CEO Clayton Williams Energy[61]

In the case of Joe Lunchbucket, there were many options in which Joe could have shared his knowledge while getting the credit for it, as it seems that was one of his concerns. Below are a few suggestions:

I. Write up an email outlining key lessons learned and send it to all drilling personnel, cc'ing supervisors and engineers.

II. Contact the drilling engineers and leadership. Suggest working together to craft an operational procedure on how to deal with the problems identified.

III. Host a morning meeting where you review the issues and solutions employed while drilling the first four Bakken horizontal wells in the new area of the play.

IV. Offer to visit all of the company rigs with or without management to share your experiences and solutions in this new area of the Bakken.

This list is not exhaustive as there are many additional options. The bottom-line is, if you are going to get up and go every day for the pay, might as well do it the right way and work as hard as possible to add value for your company. Don't be an information hoarder like Joe Lunchbucket. Share information and potential solutions. Consistently work hard and smart. As the weight-lifters like to say, "If the bar isn't bending, you are just pretending." Don't be a fake worker.

"Being busy does not always mean real work. The object of all work is production or accomplishment and to either of these ends there must be forethought, system, planning, intelligence and honest purpose, as well as perspiration. Seeming to do is not doing."

Thomas Edison,
Inventor[62]

5. KEEP: Constant Effective Communication

Excellent written and verbal communication help prevent undesirable oilfield situations. Good communication starts with inclusive pre-job planning discussions, correspondence, written procedures, team meetings, vendor meeting, and pre-spud meetings, all sufficiently before turning dirt. Once on-location activities commence, communication becomes critical to operational success and can mean the difference between life and death.

At each crew change, in addition to a handover meeting, an operational safety meeting should occur. Safety meetings provide the opportunity to communicate key aspects of the operation. Never skip a safety meeting, particularly the shift kick-off safety meeting. Furthermore, when a new task, high risk task, or non-routine operation is attempted, it is prudent to hold supplementary job-specific safety meetings. Holding an effective safety meeting is a skill in and of itself.

Safety meetings are not just a formality. A good safety meeting adds significant value. Action items should result from the discussion. Not holding a well-planned safety meeting can contribute to an undesirable oilfield situation. Many oilfield accidents, particularly fatalities, end up the subject of litigation and cite lack of a safety meeting, an ineffective safety meeting, and lack of communication as contributing factors.

It is not a coincidence that fatal accidents are often preceded by lack of communication or an issue with the safety meeting. This includes not being able to hear at the safety meeting, not having everyone involved with the operation present at the safety meeting, not discussing non-routine operations, or not discussing high risk aspects of the particular task at hand. Never assume people know the risk.

"The single biggest problem in communication is the illusion that it has taken place."
George Bernard Shaw, Playwright[63]

Poor Communication Causes Drilling Rig Fatality

The importance of excellent communication is best illustrated by studying an actual fatality in which issues surrounding communication and a safety meeting contributed to the unfortunate death of a Casing Crew Stabber. The accident occurred on a drilling rig while running casing. Since this accident is not well-known, names have been changed or removed out of respect to all parties involved. Key excerpts from the fatality investigation are below:

"The job for the day was to run casing. Prior to starting work the Rig Manager held a Pre-Tower Safety Meeting with the members of all crews. The Driller was not at the safety meeting because he was running late and not on location yet. During this safety meeting, numerous topics were discussed. One main concern expressed was the size of the tools being used on this smaller rig and the limited clearances and workspace this would create on the rig floor. . . .The concerns were answered with, *'We will just have to be careful.'*

Additional concerns were expressed during this meeting that the Casing Crew company had no job safety analysis (JSA) or standard operating procedures (SOP) for the upcoming operation. A brief JSA was written on the back of a safety booklet. This booklet could not be accounted for after the accident. The safety meeting lasted about 30 minutes after which the crews went to work.

The casing crew fielded a split crew. The reason for this was that the Blue Team casing crew had been brought in to familiarize the Red Team casing crew with equipment. . . . The two casing crews had never worked together on a casing job. During rig up, compatibility (clearance problems) were encountered between equipment due to the size of the tools versus the smaller rig floor and working space discussed in the safety meeting. . . . Rig up problems were compounded by the unfamiliarity of some of the members of the crew. Interviews with rig personnel indicated that tension, friction, intimidation, lack of understanding, and communication problems had developed between casing crew personnel.

One tool that was causing the most problems was the stabilizer. . . . A one size fits all and was much longer than it needed to be for this job. . . . The stabilizer is not specialized equipment and could have simply been shortened with a cutting torch to reduce the hazard. Prior to the accident an air hose had been snagged, hung up and broken, due to clearance problems. The casing crew supervisor on the rig floor shouted to the crew, *'Don't stop unless its broken'*. . . . As the 22nd stand was pulled through the V door, just prior to it being lifted vertically, the stabilizer hooked the bottom of the stabbing basket . . . the Casing Stabber immediately started yelling for the Driller, to stop raising the blocks. Within a few seconds, the entire crew working the floor was yelling at the Driller to stop. The Driller neither saw contact being made nor did he hear the crew yelling for him to stop. After the stabilizer hooked the bottom of the stabbing basket, its continued upward movement carried the stabbing basket up its rail guides. The Casing Stabber was wearing a full body harness tied off to the back of the derrick with a double shock absorbing lanyard system, all part of the fall protection system that he was required to wear. . . . Initially, as the stabbing basket was raised, the shock-absorbing lanyards extended. However, the extension limits were quickly reached and the nylon lanyards attached to the full body harness tightened as he was further raised. . . . At some point, the stabilizer slipped and dislodged from the bottom of the stabbing basket. With nothing holding the basket, the basket fell, shock-loading and breaking its lifting cable on the way down.

The Driller, now realizing that something was wrong, stopped the upward movement of the draw works. As the stabbing basket fell, the Casing Stabber came out of the top of the basket and fell approximately 15-20 feet. Fall protection equipment arrested his fall before he hit the rig floor. However, this was a fall of considerable distance without the assistance of the shock-absorbing lanyards, as they had already been fully extended prior to his fall. To compound the fall, he fell in a pendulum motion, slamming the left side of his body against the derrick as he came to the end of the fall.

. . . Once on the ground, the Casing Crew supervisor had him sit up and get into the front seat of his pickup with the intention of meeting the ambulance that had already been dispatched . . . the Casing Stabber could not sit in this position, so he was placed in a stretcher. The Ambulance was dispatched at 14:41, en route by 14:47, arrived on location at 15:21, and delivered him to the hospital at 15:55. While at the hospital, a chest tube was inserted to relieve trapped air within the body from a potential lung laceration; a large amount of blood was evacuated during this process due to internal bleeding… At 16:43 care was taken over by Air Recue. During the flight at 17:16, The Casing Stabber lost consciousness and was pronounced dead at 17:50."[64]

Key actions that contributed to this fatality include:[65]

1) "The Driller never went out on the rig floor to personally observe the clearances in question."
2) "The Driller's operating station was an enclosed sound proof, climate controlled room with limited visibility through a heavy glass window."
3) "The Driller was isolated from the rig floor crew and rig floor operation."
4) "The intercom system on the drilling floor was not working."
5) "The Casing Crew thought the Driller could hear them."
6) "The Driller stated that everyone assumed that he knew the clearances were close but no one ever told him directly." Remember, he missed the safety meeting when all the issues and concerns were discussed.

A number of conclusions were made regarding the fatality. The issues involving poor communication are listed on the next page.

A) "The Driller was not at the Pre-Tower Safety Meeting."

B) "The Rig Manager did not brief the Driller on what was discussed at the safety meeting that he missed."

C) "The Rig Manager did not make sure that the Driller knew of the close tolerances that he would be operating with."

D) "The Driller should have opened the door of the operator station so some voice communication from the floor could have been heard."

E) "The operator's station of this technologically advanced rig resembles a cocoon environment. His hearing is limited to what he hears through an open intercom system from the rig floor. The intercom was not working."

F) "The Casing Crew supervisor knew that the intercom was not working but he did not inform his crew."

G) "The Casing Crew did not know the driller could not hear them and was operating from their hand signals only."

H) "Safety attitude of Casing Crew supervisor, tension, lack of communication, and intimidation" contributed to the accident.

There are a number of additional issues with this operation. However, lack of communication was the central cause. The Driller missed the safety meeting, showed up late to work and didn't even bother to get out of the operating station to look over the floor before moving casing or talk to anyone. He just showed up late, sat down in his climate-controlled room, and started moving casing. Additionally, no one spoke with the Driller to bring him up to speed on the tight workspace and clearance issues. As you can see, lack of communication kills people in this business.

"One of the great failings as we look at process safety incidents is management of change. Something changed and someone forgot to tell anyone it changed. So, you have to foster an open and effective communication by ensuring those knowledge gaps do not exist."

Rex Tillerson, CEO ExxonMobil[66]

6. ABIDE: By All Governmental Laws, Rules, and Regulations

In order to follow the law, you need to know the law and how it is enforced. Therefore, understanding oil and gas regulations, which govern how laws are enforced, is key. Ignorance to rules and regulations is unacceptable and severely punished. Since laws and regulations change, staying up to date on the latest changes is a must. It also pays to understand why certain laws and regulations are in place and why they change. There is usually a very good reason. In many circumstances someone paid with their life prior to a law or regulation being passed or adopted. Let's review a legal situation regarding a company that failed to follow regulations.

Guilty of Multiple Felonies

An Oil Company pled guilty to two felony counts of violating the Outer Continental Shelf Lands Act and two felony counts of violating the Clean Water Act. According to court documents, the Company "knowingly and willfully failed to comply with the regulations for hot work on its offshore production platform. Specifically, contractors for the Company violated Title 30, Code of Federal Regulation, Section 250.113(c)(4), which mandates that welding and associated activities, also known as hot work, on offshore facilities shall not take place within 10 feet of a well bay unless production in that area is shut-in."[67]

Additionally, the Company "failed to comply with the regulations for blowout preventer testing. . . . According to the Code of Federal Regulations, the blowout preventer system must be pressure tested at regular intervals, and the entire system must pass the pressure tests

---

"Ignorance of the law excuses no man; not that all men know the law, but because it's an excuse every man will plead, and no man can tell how to refute him."

John Selden, Legal Scholar[68]

prior to resuming normal operations. According to Title 30, Code of Federal Regulations 250.617, the results of the pressure testing, including any problems or irregularities observed during the testing and the actions taken to remedy the problems, must be recorded. The blowout preventer test record and pressure chart must be signed and dated by the onsite representative as correct. The blowout preventer pressure chart and operations log are required to be maintained on the platform and available for inspection by [Bureau of Safety and Environmental Enforcement] BSEE.

According to the pressure chart for the tests conducted by contractors, only 6 of the 7 required components were tested. The chart showed pressure testing failures that required the workers on the platform to re-test the blowout preventer system. However, at the conclusion of the testing, the blowout preventer chart was not approved for accuracy by the Company on-site representative on duty nor did the workers re-test the system. . . . Then inspectors with BSEE came onboard the platform . . . for a routine inspection and requested blowout preventer testing records. The Company could not produce an acceptable pressure test chart because of the deficiencies.

. . . The Company also violated the Clean Water Act by tampering with the method of collecting the monthly overboard produced water discharge samples to be tested for oil and grease content pursuant to its [National Pollutant Discharge Elimination System] NPDES permit. The Company became suspicious that contract operators were manipulating the integrity of the overboard produced water samples at some of its platforms by filtering the sample through coffee filters or other similar means to ensure that the Company would not be found to be in violation of its Permit. Although the Discharge Monitoring Reports for the platforms . . . showed that the platforms were not discharging oil and grease in excess of the permit requirements, when the Company began an investigation, the results of which it self-reported to the United States, and took the samples in accordance with the Permit requirements, multiple platforms were shown to be in violation of the monthly discharge allowances." [69]

Recently "two contract operators on the Company's oil production platform were engaged in bleeding pressure from the production casing on a plugged well. Operators routinely encounter liquid, including pollutants such as wellbore fluid, acid, and hydrocarbon/oil residue, when bleeding pressure from well casings, and therefore precautions against an unpermitted discharge should be taken. . . . The two contract operators onboard did not take any precaution against a discharge of pollutants when they began bleeding down the pressure from the production casing. The contract operators attached a hose to the valve from which the casing pressure was to be released and put the end of the hose at the edge of the platform, allowing wellbore fluid mixed with hydrocarbons to shoot out falling into the water below in violation of the Clean Water Act."[70]

Operations Analysis

There is no good reason to violate oil and gas regulations. Although this business spans the globe, often taking place in remote areas, everyone seems to know each other. It's a small industry. Once you get a reputation for committing oilfield crimes, it will impact your career much further than you think. You could also end up in prison. It's not worth it. If a customer asks you to do something, do not assume they know the regulations. Just because they are a boss does not automatically make them more knowledgeable than you. They may not understand what they are asking. If something sounds wrong to you, don't do it.

"Regard your good name as the richest jewel you can possibly be possessed of – for credit is like fire; when once you have kindled it you may easily preserve it, but if you once extinguish it, you will find it an arduous task to rekindle it again. The way to a good reputation is to endeavor to be what you desire to appear."

Socrates, Philosopher[71]

7. ALWAYS: Have a Written Procedure

Great power exists in the written word. When everything is gone from location, a wellbore and written words remain, evidence of hard work on Earth. Some say it's a measure of our progress in the universe, objective evidence of what separates intellectually advanced man from Cro-Magnon Man, the oldest known modern man. Others say the oilfield is full of Neanderthals or animals roaming from location to location in human form. Truth is, we are all animals, and just like our less advanced animal friends, we all want to be free, think free, live free, unconstrained by cages and chains, or bureaucratic procedures and stifling process for us, oilfield hands.

It does not have to be that way. Oilfield procedures can be liberating to all those who venture into the unknown. If you have a clear procedure, you can set your mind free to focus on the task at hand and not worry that you are forgetting something. A good procedure also reduces implementation anxiety by providing a roadmap. If you need to get from Point A to Point B, not having a roadmap makes the journey much more difficult and dangerous.

No Procedure Used As Evidence Of Negligence

The importance of having a procedure is best illustrated by studying a court case in which lawyers use lack of procedure and process to argue negligence regarding a near fatal injury. The case made its way through the legal system up to the Supreme Court of Oklahoma and was decided in February 2010. Names have been changed or removed out of respect to all parties involved. Key excerpts (abridged and adjusted) from legal documents are summarized below:

"Arnold was severely injured while working as an independent contractor providing bulldozing services to a Drilling Company. . . .

The Drilling Company owned the drilling rig and its various components and had contracted with Conan Crews to move all the components from an old drilling site to the new one. Conan Crews subcontracted Wizard Trucking for the use of a tandem truck to move the rig and for the services of the owner and operator.

The job required moving two sets of rig substructures, each consisting of a bottom 'pony' sub and an upper 'top' sub and stacking the top subs onto the pony subs. These substructures are massive iron pieces which must be strong enough to support the derrick and other rig components; they are approximately 47 feet long, 11 feet tall and weigh more than 50,000 pounds each. At the time of the accident, Arnold was the operator of the bulldozer and was assisting a forklift driver and Tandem Truck operator in an effort to place a top sub onto a pony sub when the top sub toppled off the pony sub and fell onto the cab of the bulldozer, crushing the cab and severely injuring Arnold.

Arnold brought suit in negligence for damages against the independent contractors, Conan Crews and Wizard Trucking, and also against the Drilling Company. Conan and Wizard settled; however, Arnold proceeded to a jury trial against the Drilling Company and presented evidence to show that they had control of all activities at the worksite; owned, maintained, and controlled all the rig components, including the top sub and the pony sub involved in the accident; and failed to adequately plan, supervise and manage the rig-up procedure.

Evidence, though challenged, includes the following:

1) The Drilling Company did not include the rig-up procedure in its safety manual and did not have a safety meeting regarding the process.

2) The Toolpusher was the overall supervisor and had ultimate authority, the 'final say' on all matters including safety.

3) On the day of the accident, the Toolpusher was not present to supervise the operation even though stacking the subs was the most dangerous part of the rig-up procedure.

4) The 'relief' Toolpusher on duty that day was unfamiliar with the rig, the location and the people involved, and was not present during the stacking effort.

5) Though aware that the safest method to stack the subs was by using a crane, the Drilling Company did not require or allow for the expense of a crane.

6) The Drilling Company had provided a location which was too small and cramped for the normal procedure of using two gin trucks to lift the top sub from the tandem truck with their winches.

7) The derrick had been assembled too close to the location, so there was not adequate space for the trucks to work, necessitating that smaller vehicles, the forklift and Arnold's bulldozer, were used instead.

8) The pony sub was an unsafe instrumentality because it did not have full-length channel iron guides which would have prevented the top sub from falling.

9) The top runners of the pony sub should have been cleaned and lubricated.

The last two points were offered to show that the Drilling Company should have altered the pony sub by preparing and equipping it with full-length channel guides into which the bottom rails of the top sub could have been placed in order to safely slide the top sub into a stable position.

4,821 workers died on the job in the U.S. in 2014, the highest since 2008

Bureau of Labor Statistics[73]

Testimony was presented to show this modification could have been accomplished easily and inexpensively and would have prevented the accident. The court held that the accident would not have occurred if:

A) There had been full channel guides on the pony subs.
B) A crane had been used.
C) The Drilling Company had the foresight to clean and lubricate the top runners on the pony sub.
D) The top sub hadn't gone a little bit crooked and perhaps hit a guide and kicked up on the sub and then slid off.
E) The Toolpusher had been there and had a safety meeting before they started and said, *'Here is the way we need to put those subs together.'* "[74]

The jury ruled in favor of Arnold over the Drilling Company. The verdict was overturned by the Court of Appeals, but reinstated and affirmed by the Supreme Court of Oklahoma. The entire legal process took nearly 9 years from the time of injury.[75]

Operations Analysis

The foundation of a low-risk operation is a written procedure, particularly for the most dangerous aspects of the operation. In this case, it was the stacking of the substructure. If a thorough detailed task-specific, step-by-step procedure was available to all personnel on location, reviewed during the safety meeting and implemented with oversight, the issues listed above would likely have been addressed.

One procedural tool to consider utilizing is a Job Safety Analysis (JSA), which covers the sequence of job steps, associated potential hazards, and recommended safe procedures, often written in a table format. My opinion is that a JSA should have been utilized and would have added significant value in helping prevent this accident.

"The great obstacle to discovering the shape of the Earth, the continents, and the ocean was not ignorance but the illusion of knowledge."
Daniel J. Boorstin, Historian[76]

8. DRIVE: Defensively

Statistically, transportation accidents are responsible for the majority of fatalities in the oilfield. Since oil and gas well locations are spread across great distances, and we have to travel to location to work, we are exposed to a significant amount of travel time and transportation risk. In many cases, the element of fatigue plays a part. We drive to location, work a full day on location, then drive home. The journey and associated risk must be managed.

Based on 2014 data from the U.S. Department of Labor, Bureau of Labor Statistics, the most recent data available at the time of publishing, 49% of oil and gas fatalities (69 of 142 fatalities in yr-2014) are transportation related, primarily roadway incidents involving vehicles.[77]

2014 Oil And Gas Fatality Statistics[78]

Fatality Classification	# of Deaths	% of Total
Transportation	69	49%
Struck by Object	32	23%
Fires and Explosion	17	12%
Falls, Slips, Trips	12	8%
Exposure	6	4%
Other	6	4%
Total	**142**	**100%**

Note: Based on NAICS codes 211, 213111, 213112.
2014 BLS data considered preliminary at the time of research.

2014 is not an anomaly. On average since 2009, over 40% of fatalities in the oil and gas industry are transportation related, mostly involving motor vehicles on roadways.[79]

"Highway vehicle crashes are the leading cause of oil and gas extraction worker fatalities. Roughly 4 of every 10 workers killed on the job in oil and gas are killed as a result of a highway vehicle incident."

OSHA[80]

2009 - 2014 Oil And Gas Fatality Statistics[81]

Fatality Classification	2009	2010	2011	2012	2013	2014
Transportation	40%	38%	46%	46%	39%	49%
Struck by Object	31%	19%	23%	17%	22%	23%
Fires and Explosion	7%	23%	11%	17%	12%	12%
Falls, Slips, Trips	-	7%	8%	13%	13%	8%
Exposure	12%	8%	8%	6%	7%	4%
Other	10%	5%	4%	1%	7%	4%
Total # of Deaths	**68**	**107**	**112**	**139**	**112**	**142**

Note: Based on NAICS codes 211, 213111, 213112. 2014 BLS data considered preliminary at the time of research.

Reduce transportation risk by employing good Journey Management. One tactic to consider is to build a personalized travel checklist and review it before each trip. Incorporate things you might forget, aspects regarding your vehicle, road risks, and work environment. Below is an example.

Oilfield Road Warrior Rules Of Survival

1) Know Yourself → Fatigue, Distractions
2) Know the Vehicle → Blind Spots, Weaknesses, Strengths
3) Know the Road → Study and Plan Route Before the Trip
4) Know the Weather → Impact on the Journey
5) Walk Around the Vehicle Before Getting In → Check 360°
6) Secure All Loads → Including Your Body; Buckle Up
7) Use a Spotter Before Moving It → Maintain Communication
8) Keep Your Distance From Others → People Can Be Crazy
9) Always Have An Escape Plan → Never Get Trapped
10) Drive With Honor

Globally, 1.2 million people die in road crashes and
50 million are injured in road crashes each year.
ASIRT[82]

Dog Fatality Destroys Oil Man And Company

One of the most heartbreaking stories in this book involves a Company Man named Max and his dog, Lord Hugo. Max worked in Australia for Seven Sisters Resources, a company which did not mind if an employee kept a dog on location. Lord Hugo, a friendly 175 pound Great Dane, would sit in the shade around location in an effort to escape the hot Australian sun. Sometimes he would sit underneath pickup trucks.

One day the owner of Seven Sisters, Henry Mattei, visited location in his brand new dually pickup. Max and Lord Hugo greeted him and they toured the location together. At some point, Lord Hugo wondered off. At the end of the visit Henry, who was running late for a meeting, jumped in his pickup and left in a hurry, barreling off location. What he did not know was that Lord Hugo was laying under his pickup truck. When Henry took off, Lord Hugo could not get up in time and got wedged in between the rear wheels, crushed and dragged until parts of his body were scattered across location.

Max witnessed the tragedy unfold in complete shock. Henry had no idea and continued to drive away. Max ran to his pickup and proceeded to chase down Henry. Once he caught up to him, Max forced Henry into the ditch. Max, in tears and completely out of his mind, attacked Henry until he was severely beaten. Eventually, Max was arrested by the police.

Once Henry realized what he did, he was devastated. However, he fired Max anyway. Henry also pressed charges, which resulted in 12 months of jail time for Max. Seven Sisters Resources and Henry had a very hard time finding anyone to work for them as word of what Henry did quickly made its way across the oil patch. Ultimately, Henry had to sell the company because no one would work for him.

"If you don't have a dog - at least one - there is not necessarily anything wrong with you, but there may be something wrong with your life."

Vincent van Gogh, Painter[83]

Remember, in the oil and gas business, your reputation is everything. Max, after being released from prison, fell into a deep depression. His desire for life faded away, all that remained were memories, as Lord Hugo was his best friend and he felt responsible for what happened. He was never the same after the loss of Hugo and his time in prison.

Case Analysis

Before you move your vehicle, think of the tragedy of Lord Hugo, Max, Henry, and Seven Sisters Resources. Never move your vehicle before walking around it (Check 360°). Inspect all equipment to ensure its secure. LOOK UNDERNEATH to make sure you don't run over anything or anyone. You do not want to have to live with this type of tragedy for the rest of your life. When you Check 360°, you honor the great spirit of Lord Hugo, empathize with the mental anguish of Max, and give tribute to the destruction of Seven Sisters.

STATISTICS: UNITED STATES ROAD CRASHES

- Over 30,000 people die in road crashes each year.[84, 85]
- 28% of fatalities are speeding related (yr-2014).[86]
- 21% of fatalities are fatigue related (drowsy drivers).[87]
- 31% of fatalities are alcohol related (yr-2014).[88]
- 10% of fatalities are distracted driving related.[89]
- 2,300,000 people are injured or disabled in crashes per year.[90]
- Seat belts save over 12,000 lives per year.[91]
- Road crashes cost the U.S. $230 billion per year.[92]
- Oil & Gas vehicle fatality rate is 8x that of all industries.[93]
- 51% of Oil & Gas vehicle fatalities occur in pickup trucks.[94]
- 27% of Oil & Gas vehicle fatalities occur in tractor trailers.[95]

9. MAINTAIN: Situational Awareness

From a certain point of view, working in oil and gas is like going to war—things can kill you from any and all directions. The oilfield warrior must maintain diligence and awareness at all times. Oilfield bullets can and will fly at you when you least expect it. Death from above due to falling objects, death from behind when you are not alert, death from the sides, and even death from the dirt as the following court case demonstrates. Since this accident is not well-known, names have been changed or removed out of respect to all parties involved.

Lack of Awareness Results In Workover Fatality

"The workover job was to drill out a cast iron bridge plug and clean out to bottom on a Tuscany Oil Company well. Pisa Tower Energy Services set up their workover rig on November 15, and until December 5 of that same year, the workover operation proceeded without incident. . . . It rained on December 5, and continued raining on December 6, the day of the accident. On December 5, the crew began pulling out of the hole for the fourth time, but had only removed 30 stands before they stopped for the day.

. . . The next day they continued. . . . While pulling pipe, Tony, the Toolpusher, notified Mr. Caine, the Tuscany Company Man, that the racking board was sinking. Mr. Caine stayed in his truck and did not check out the situation. Instead, he relied upon Tony's assurance that it was nothing to be concerned about.

. . . Later that day, the derrick man, Darren, told Giovanni (the driller) that he thought that the derrick was leaning. Giovanni testified that he performed a test that indicated that the derrick was level, and decided to proceed. . . . They continued pulling pipe . . . up until the workover rig and derrick fell over, killing Darren and injuring Giovanni. There were 236 stands of pipe racked in the derrick, with an estimated weight of 135,000 pounds. Mr. Caine, the Company Man, was at the site waiting in his truck for the crew to complete the procedure, so he could inspect the drill bit.

Tuscany Oil and Mr. Caine argued that they are not liable for the death and injuries of the employees of Pisa Tower Energy Services because:

(A) Tuscany and Mr. Caine were not owners of the premises.

(B) There is no duty to protect an independent contractor against risks arising from or intimately connected with the work, and there is no liability for death or injury resulting from dangers that the contractor, as an expert, knows or should know.

(C) There was no duty to perform a soil test or site preparation.

The court ruled that Tuscany Oil and Mr. Caine were responsible for the following reasons:

While Tuscany Oil was the lessee and not the owner of the land, it clearly was the owner of the well, and it admitted so on several occasions. When Mr. Caine, Tuscany's Company Man, was being cross-examined, he admitted that Tuscany Oil Company owned the well in the exchange that follows:

Q. 'Okay. Now did Tuscany own that well on December 6?'

A. 'Yes.'

Q. 'Did they own that well on November 15, when the workover rig was pulled back on to the well site by a wrecker?'

A. 'They owned it then, yes, sir.'

Pisa Tower Energy Services crew was required to follow Tuscany's procedure, under the supervision of Tuscany's Company Man who exercised control over the down hole operations and any related work. . . . Pisa Tower Energy Services crew was following Mr. Caine's order to continue pulling pipe until all of the pipe was pulled before the end of the day.

. . . Other testimony at trial confirms these assertions. Tuscany Oil Company admitted that it directly supervised the down hole operation. Further, when Mr. Caine, the Company Man, was being cross-examined concerning the workover plan prepared by Tuscany that Pisa Tower Energy Services followed, this exchange took place:

Q. 'Well, they had to follow that step by step procedure there, didn't they?'

A. 'They had to - we have to have a beginning point and we have an end result and these are the steps in between and we try to get to that point.'

Q. 'And who is we?'

A. 'The prognosis is written by Tuscany and this work and each step is carried out by Pisa Tower Services and I don't - don't physically go out there and do any of that work myself. That's the Pisa Tower Services crew and they are responsible for the safety and the operation of that rig while they are doing what we direct them to do from this prognosis.'

When Bruno Bonanno, the other Company Man and Mr. Caine's relief, was being cross-examined, he admitted that Tuscany Oil Company had the ultimate control over the workover: 'I do have control over what - when they stop and start.' He further admitted he was the 'boss' over the workover site:

Q. 'You can stop that rig and shut it down in the face of the Tool Pusher's objection if you deem it necessary for safety, can't you?'

A. 'Yes, sir, I could.'

Q. 'Because one more time, you are the boss?'

A. 'Yes, sir.'

"Never underestimate the power of a well-framed question."
Lee Tillman, CEO Marathon Oil[96]

> *Q. 'Okay. Then I'll rephrase my question and ask you that you or whichever Company Man is out there that day is the boss no ifs, ands, or buts about it?'*
>
> . . .
>
> *Q. 'Do you understand that question?'*
>
> *A. 'Yes, sir.'*
>
> *Q. 'Okay. And that's the truth, isn't it? That's what you are out there for, isn't it?'*
>
> *A. 'Yes, sir.'*
>
> *Q. 'So, there's no question about that?'*
>
> *A. 'No, sir.'*
>
> *Q. 'Okay. You can make them stop any time you want to if you feel that's necessary?'*
>
> *A. 'Yes, sir, if I feel it's necessary.'*
>
> *Q. 'You can make them work any time you want to if you think that's necessary?'*
>
> *A. 'Yes, sir.'*

Giovanni's (the driller) testimony also indicates that the ultimate control of the workover site was in the hands of the Company Man, who was on the job site 75-85% of the time:

> *'Well, the Company Man, first of all, he don't tell me nothing. He goes to my Tool Pusher. My Tool Pusher tells me. I tell the crew that's under me. That's all work in the chain of command. Anything that I got to ask. I don't go to the Company Man and ask. I go to my Tool Pusher. My Tool Pusher goes to the Company Man. The company man goes to call whoever he's got to call.'*

When another employee of Pisa Tower Energy Services was asked whether Tuscany's Company Man had the right to terminate his personnel for any reason, he responded, *'He could - he could stop operation of the rig and lay the rig off.'*

Although an owner is not liable for injuries sustained by an employee of an independent contractor caused by the negligence of the independent contractor, an owner is liable to employees of an independent contractor for his own negligence. Mr. Caine and Tuscany Oil similarly had the authority to stop any work not being performed safely. Testimony demonstrated that Tuscany did retain and exercise substantial control over Pisa Tower Services work and directly supervised the down hole operation.

A short time before the rig fell over, Tony the Toolpusher had notified Mr. Caine the Company Man that the racking board was sinking. This presents a jury question, and a jury could reasonably find Tuscany and Mr. Caine negligent for either failing to recognize their responsibility, or recognizing it but failing to specifically instruct the workers.

While the general rule is that the owner of the premises does not have a duty to protect an independent contractor against risks arising from or intimately connected with the work, there is an exception where the owner maintains substantial legal and factual control over the work to be performed.

Tuscany Oil Company and Mr. Caine next argued that the contract charged Pisa Tower Energy Services with knowledge of the site conditions, and Pisa Tower Energy Services employees were in a superior position to observe the response of the soil and site conditions to their work. They argued that Tony, the Toolpusher, testified that the site was soft and that is why Pisa Tower Energy Services decided to use a double wide beam to support the workover rig. Also, Giovanni and Tony had observed the sinking of the pipe mat on which they racked the pipes several days before the accident and decided to add additional boards for support. Tuscany and Caine contended that they had no duty to warn of a danger Pisa Tower and its employees should reasonably have appreciated.

. . . However, Tuscany was not just a detached owner who hired an independent contractor to perform some work. Tuscany provided Pisa Tower Energy Services with a prognosis that detailed, step-by-step, how to perform the work. Tuscany had a Company Man at the work site and retained substantial control over the job. Tuscany's District Production Manager admitted that it had a duty to provide a safe work site in the following exchange:

> *Q. 'Did Tuscany have a duty or responsibility to provide a safe place to work for these men that you hired to come out there and rig up on Tuscany's property? Did you have that duty?'*
>
> *A. 'It's our - it's our duty and this is what we try to do in all cases to provide as safe working conditions as we can for anybody. We are not going to let our thoughts about profits or anything override our concern for safety in any case that we see that we can do something about. That is our policy.'*

Therefore, Tuscany had a duty not only to warn of dangerous conditions, it had a duty to supervise in a safe manner.

Tuscany and Mr. Caine next argued that there was no basis in the record for finding a duty to perform a soil test or any additional site preparation; no evidence of a dangerous soil condition; and no evidence of a causal connection between any alleged failure and the accident. According to Tuscany, all witnesses who offered testimony on the issue of site preparation testified that, whether by custom of the industry, or by the course of performance of the parties, Pisa Tower Services had the responsibility to decide what, if any, additional site work was needed in order to perform its work safely. If such site preparation was necessary, Pisa Tower Energy Services would request Tuscany to perform such work before Pisa Tower Energy Services would rig up its equipment. Further, both tool pushers testified that the site at issue was sufficiently prepared to perform their work safely, without any additional site preparation, and there was no evidence of any dangerous condition. However, every witness who testified concerning this issue stated that Tuscany was responsible for the well site preparation. Further, it is undisputed that Tuscany did nothing to inspect or prepare the well site before Pisa Tower Services began performing the workover. Mr. Caine agreed that it was absolutely necessary that the pipe rack be supported by a suitable foundation.

When Tuscany's District Production Manager was being cross-examined at trial, the following exchange took place:

Q. 'Well, would you agree with me that at least one of the safety factors that you would look at would be the condition of the soil underneath the racking board? That would be at least one factor that could have caused this rig to fall over, could it not?'

A. 'That could be at least one factor but not the sole factor.'

Q. 'All, right. And if that were a factor and if it's your duty as Tuscany to provide a safe place to work, would one of those duties in providing a safe place to work be site preparation?'

A. 'Yes, one of those duties would be site preparation work.'

Q. 'And there's no duty whatsoever upon Pisa Tower or any of their employees to do any site preparation work, is it?'

A. 'There is a duty upon everyone when it comes to safety.'

Q. 'That wasn't my question. My question was whether or not they had any duty with reference to site preparation work?'

A. 'They have no duty to my knowledge toward - toward site preparation work.'

Q. 'Well, did Tuscany hire a soil expert before they rigged up on this location?'

A. 'Not to my knowledge.'

Q. 'Did Tuscany prep this site like it would have been prepped a site where they were drilling a new well, did they do that?'

A. 'Not to my knowledge.'

Dr. Stone was offered as an expert in the field of soil mechanics as a civil engineer who was able to give an opinion as to why the rig overturned. . . . He testified that if crushed stone or gravel had been added to the site prior to commencing the workover, the soil would have been sufficiently strong to support the weight of the pipes and the accident would not have occurred."[97]

Operations Analysis

There are a number of learnings from this unfortunate accident and fatality to review before moving forward. Before the rig fell over, at least two people noticed that there was an issue and notified their supervisors. No action was taken to address the situation or shut the job down.

Tony, the Toolpusher, told the Company Man, that the racking board was sinking. Mr. Caine (Company Man) stayed in his truck and did not check out the situation. There is no excuse for this. If someone brings an issue to you, regardless of your job, you are obligated to look into it. The Toolpusher would not bring up an issue if he was not concerned about it. If you are a company man and a service provider mentions an issue, a red flag should go up in your mind. Ignoring something will not make it go away. You must take action and look into it in great detail. Do not look for a reason to dismiss the concern.

Additionally, around thirty minutes before the rig turned over, the Derrick Man, Darren, told Giovanni (the Driller) that he thought that the derrick was leaning. Giovanni testified that he checked whether it was level and did not see an issue. However, the derrick was leaning but no one shut the job down or looked further into the situation. This is poor situational awareness by leadership on location.

Anyone can shut the job down, but nobody did. From the Derrick Man to the Company Man, nobody exercised Stop Work Authority to shut the job down. Clearly, Tony was reaching out to the Company

Man regarding the racking board sinking. When the Company Man did not even get out of his truck to take a look, it demonstrates lack of concern. Therefore, Tony probably felt pressured to continue working. Tony should have stopped the job and addressed the situation. It is difficult when the Company Man does not seem to care. However, Tony still should have addressed the situation. Although the Company Man is ultimately the boss, everyone on location has Stop Work Authority.

Darren, the Derrick Man, knew the rig was leaning and should have also stopped the operation. He did tell his boss, Giovanni the Driller, who inaccurately assessed that the rig was level and told Darren everything was good. In my experience, when something is not going right, there are warning signs that are often ignored and explained away. People don't want to have problems and never think something bad will happen to them. As humans, we often rationalize negative issues as being insignificant, ignore them, and hope they go away or are not really issues that need to be addressed. In some fashion, this aspect of human nature probably played a part in this accident.

The weather also contributed. Always be aware of how weather can affect an operation. Heavy rain definitely will impact an oilfield operation in many ways. No one even thinks about the stability of the ground they are standing on, but as this accident shows, we must always maintain situational awareness from every direction.

"It's not what you look at that matters it's what you see."

Henry David Thoreau, Poet[98]

In summary:

1) Maintain situational awareness.
2) Monitor the weather and its impact on operations.
3) Perform a thorough site inspection every day.
4) Ensure pad stability prior to and during operations.
5) Rack pipe on a stable suitable mat for the operation at hand.
6) Take concerns seriously and look into them in detail.
7) Never ignore data that indicates you have a problem.
8) Never dismiss warning signs, especially when expressed from people on location during an operation.
9) Never hesitate to shut a job down.
10) Everyone in the oilfield has the right and obligation to Stop Work if anything is unsafe or there are concerns.

"Ladies and gentlemen, I do my own drillin' and the fellers that work for me are fellers I know. I make it my business to be there and see to their work. I don't lose my tools in the hole, and spend months a-fishin'; I don't botch the cementin' off, and let water into the hole, and ruin the whole lease... I assure you whatever the others promise to do, when it comes to the showdown, they won't be there."

Upton Sinclair, Oil[99]

10. STAY: Disciplined

It is easy to be consistently focused on your work every minute during critical operations for a few days or a few weeks. However, maintaining the discipline to focus on critical tasks every day for years is difficult. Although disciplined focus is challenging, this is what is required, as the oil and gas business demands our full attention or incidents will occur. If you take your eyes off the "dragon," you will get burned in this business. The area where this rule is most imperative is the business of well control.

To illustrate this point, a select list of blowout and well control incidents, with a focus on recent events, is presented. The data is sourced from the Railroad Commission of Texas. Based on available information, each incident is grouped into one of eight categories for statistical analysis. As you read through the list, think about how each one of these situations could have been prevented. It helps to review the list regularly to remind yourself of all the different ways loss of well control incidents occur.

Well Control Incident Tables[100]

Year	Classification	Loss of Well Control Incident
1990	Workover	Workover blowout when pulling sucker rods
1990	Drilling	Underground blowout with surface cratering
1991	Production	Production tree failure
1991	Plugging	Incident during plugging operation
1991	Production	Wellhead separated from casing
1991	Collision	Truck backed over production tree
1992	Production	Packer failed on producing well
1992	Collision	Tree chopper hit well
1993	Production	Tubing valve failed
1993	Frac or Flowback	Incident during fracture stimulation
1993	Collision	Backhoe hit well
1993	Production	Downhole packer failed

Year	Classification	Loss of Well Control Incident
1994	Drilling	Drill Stem Test well control incident
1995	Workover	Incident during recompletion
1995	Drilling	Well control problem when changing bit
1995	Drilling	Pulling drill pipe when well came in
1995	Workover	Loss of control when drilling out bridge plug
1995	Frac or Flowback	Frac job well control incident
1996	Workover	Incident during routine pump change
1996	Production	Well control situation on producing well
1997	Workover	Changing out ESP equipment incident
1998	Workover	Casing split during workover
1998	Drilling	Blowout during bit trip
1998	Wireline	Blowout during logging operation
1998	Drilling	Lost circulation led to well control incident
1998	BOP Incident	Installing BOP when problem occurred
1998	Drilling	Incident when pulling core barrel out of hole
1999	BOP Incident	Removed BOP when incident occurred
1999	BOP Incident	Annular preventer leak
2000	Collision	Tractor sheared off wellhead
2000	Drilling	Blowout during drilling
2000	Drilling	Well control incident when running casing
2000	Drilling	Hole in casing during drilling
2000	Production	Leak on producing well
2000	Drilling	Blowout while tripping out of well
2001	Workover	Swabbing well when kick occurred
2001	Production	Flowing gas between casing and tubing
2001	Drilling	Bull plug failure during drilling
2001	Workover	Well kicked when pulling tubing
2003	Workover	Blew last 3 tubing joints out of hole
2003	Production	Choke valve cut out on production tree
2004	Drilling	Wellhead gave up directly below BOP
2004	Drilling	Lost circulation, blowout ignited at shaker
2004	Production	Flowline split incident

Year	Classification	Loss of Well Control Incident
2005	Frac or Flowback	Production casing rose-up out of the ground
2005	BOP Incident	Unable to close BOP's
2005	Production	Grease fitting failed on tree
2005	Frac or Flowback	Casing parted during fracture stimulation
2005	Production	Closed valve did not hold
2005	Plugging	Incident during plugging operation
2005	BOP Incident	BOP seal started leaking
2005	Production	Blowout through tubing/casing packing
2006	Production	Leaking gas through needle valve
2006	Production	Valve failure on tubing/casing annulus
2006	Production	Broken casing wing valve incident
2006	Collision	Truck backed over injection well
2006	Production	Unable to close casing valve
2006	Wireline	Pulling wireline out of well when well kicked
2006	Drilling	Hole in bradenhead
2006	Drilling	Incident when tripping out of hole
2006	Drilling	Making bit trip when well came in
2006	Workover	Snubbing equipment failed during operation
2006	Drilling	Well kicked during connection
2006	Collision	Truck backed into well
2007	BOP Incident	Incident when removing BOP's
2007	Frac or Flowback	Tubing parted, then casing failed
2007	Frac or Flowback	Wellhead equipment failure
2007	Frac or Flowback	Wellhead failure below frac valves
2007	Drilling	Port on wellhead failed below BOP
2007	Drilling	Tripping drill pipe, then well control incident
2007	Production	Lower valve failed when changing upper valve
2008	Frac or Flowback	Frac stack valve failed
2008	Collision	Landowner knocked valve off with dozer
2008	Drilling	Bradenhead cracked causing incident
2008	Collision	Vehicle ran over wellhead causing blowout
2008	Drilling	Well control issue, closed BOPs, then blowout

Year	Classification	Loss of Well Control Incident
2008	Drilling	Surface casing failure
2008	Drilling	Blowout during bit trip
2009	Production	Casing valve cut out while bleeding pressure
2009	Frac or Flowback	Loss of well control after frac job
2009	Drilling	Incident during fishing when tripping out
2009	Production	Gas flow between casing flange & master valve
2009	Workover	Snubbing out packer when tubing parted
2009	BOP Incident	Well assumed dead, BOP removed, blowout
2010	Wireline	Blowout during logging
2010	BOP Incident	Packer got stuck in BOP, then blowout
2010	Drilling	Tripping drill pipe when incident occurred
2010	BOP Incident	Pipe was cut, well began to flow, BOPs failed
2010	BOP Incident	Drilled out plug, well came in, BOPs failed
2010	Wireline	Perf well, expected oil but got gas, then blowout
2010	Drilling	Drilling 50' from old well when old well blew
2010	Drilling	Kick during drilling, BOPs failed
2011	Workover	Drilled out cement plug when well came in
2011	Production	Tubing parted causing incident
2011	Drilling	Tripping in drill pipe, well started unloading
2011	Drilling	Hit gas pocket during drilling
2011	Frac or Flowback	During frac flowback Weco connection failed
2011	Frac or Flowback	Casing ruptured during frac job
2011	Workover	Loss on control during workover operation
2011	BOP Incident	BOP removed to install wellhead, well blew out
2011	Drilling	Started flowing when running surface casing
2011	Production	Production tubing sanded up and blew out
2011	Frac or Flowback	Casing parted during frac, wellhead separated
2011	Wireline	After perforating, well blew out
2011	Workover	Loss of well control during workover
2011	Drilling	Blowout while tripping in new bit
2012	Production	Split in tubing below stuffing box
2012	Drilling	Set surface casing, lost circulation, lost control

Year	Classification	Loss of Well Control Incident
2012	Workover	Pulling rods when well blew out
2012	Frac or Flowback	Lost control during frac job
2012	Wireline	Well came in after wireline run
2012	Plugging	During plugging operations, blowout occurred
2013	Production	Valve failure caused well control incident
2013	Production	Valve accidentally left open in cellar
2013	Frac or Flowback	Frac head failed during frac job
2013	Workover	Well started flowing during ESP removal
2013	Workover	Lost control when pulling tubing
2013	Frac or Flowback	Production casing parted during frac job
2013	Workover	Blowout during swabbing
2013	Workover	Blowout during ESP installation
2013	Wireline	Blowout during slickline deviation survey run
2013	Frac or Flowback	Drilling out frac plugs when incident occurred
2013	Production	Packing gland at wellhead failed
2013	Frac or Flowback	Blowout while flowing back gas and sand
2013	Frac or Flowback	Casing parted during frac job
2013	BOP Incident	Cemented casing, removed BOP, blowout
2013	Production	Rubber packing around tubing failed
2013	Frac or Flowback	Well started blowing while fracturing offset well
2013	BOP Incident	Cemented well, Unbolted BOP, blowout
2013	Production	Needle valve on tree cut out
2013	Frac or Flowback	During flowback, flange failed on flowline
2013	Drilling	Well came in during wiper trip
2013	Drilling	Kick, BOP shut, fluid at surface 100' from well
2014	Drilling	On 2nd stage of cement job, well began flowing
2014	Workover	Sealing element failure during snubbing ops
2014	Workover	Loss of control during workover
2014	Drilling	Kick when POOH, gas reached diesel engine
2014	Frac or Flowback	During frac, fluid started flowing from cellar
2014	Drilling	Blowout during cement job
2014	Drilling	Loss of surface casing integrity

Year	Classification	Loss of Well Control Incident
2014	BOP Incident	Removed BOP and well began to flow
2014	Plugging	Well control incident during plugging job
2014	BOP Incident	Installed BOP, rubbers failed overnight
2014	BOP Incident	BOP did not shut properly
2014	Plugging	During plugging and pulling joints, blowout
2014	Workover	Well control situation during workover
2014	Frac or Flowback	5.5”x7” communication on frac, wellhead failed
2014	Frac or Flowback	7” casing parted 10’ below stack during frac job
2014	Production	Casing valve left open in cellar
2014	Production	Wellhead cracked
2014	Frac or Flowback	Incident during frac plug drill out, BOP failed
2014	Frac or Flowback	Loss of control during frac job flowback
2015	Drilling	Blowout while drilling surface hole
2015	Frac or Flowback	Gasket started leaking during frac job
2015	Frac or Flowback	Blowout during frac job
2015	Frac or Flowback	Leak below master valve during frac job
2015	Workover	Loss of well control when running tubing
2015	Production	Pinhole leak caused incident
2015	Frac or Flowback	Intermediate casing ruptured during frac job
2015	Drilling	Gas leak from surface during drilling operations
2015	Production	Producing well loss of control
2015	Workover	Well control situation on workover operation

"If you don't pay a guy enough to drill your well, it can cost you a helluva lot more in the long run, because that driller's insurance just says there have to be blowout preventers on the wellhead. It doesn't say whether they have to be working or not. You have to make damn sure you test it. That's the reason you have blowout drills. And you have to check the mud weights often, and check the temperature and what's in the mud. Is there any gas coming up with it?

Old-time drillers used to check their cuttings all the time.
...Nowadays, drillers don't do that.
They find someone to do it for them.
They hire consultants with computers.
Then they get in trouble."

Red Adair, Oilfield Firefighter[101]

Blowout Risk Statistics 1990 - 2015

Segmenting the " *Well Control Incident Tables* " list of well control incidents into eight categories, we can see which operations carry a higher propensity for a blowout to occur. When looking at the data from 1990 to 2015, it comes as no surprise that blowouts during drilling account for the highest percentage of total blowouts. Most folks associate blowouts with drilling activities. However, production, fracturing, and workover, in conjunction, account for 53% of total blowouts.

Rank	Blowout During	Percent of Total Blowouts
1	Drilling	25%
2	Production	20%
3	Fracturing or Flowback	18%
4	Workover	15%
5	BOP Removal or Incident	10%
6	Vehicle Collision	5%
7	Wireline Operation	4%
8	Plug & Abandon	3%

Drilling blowouts include the loss of well control during activities with a drilling rig on location, including the following operations:

- Drilling
- Coring
- Running casing
- Cementing
- Tripping
- Fishing
- DST
- Casing failure during drilling
- Wellhead failure during drilling

TEST YOUR SKILLS

What drilling activity has the highest propensity for a well control situation?

Statistically, with this dataset, 37% of drilling blowouts occur during tripping; the highest percentage within drilling activities.

Production blowouts include loss of well control during production related activities, including the following:

- Production tree failure
- Wellhead failure
- Downhole packer failure
- Annular gas flow
- Flowline failure
- Valve failure
- Packing failure
- Tubing part
- Open valve (forgot to close valve)

Fracturing blowouts include the loss of well control during frac job related operations due to:

- Casing failure
- Tubing failure
- Wellhead failure
- Frac stack failure
- Offset well frac hit
- Flowback iron failure
- Hammer union failure
- Frac iron failure

TEST YOUR SKILLS

During fracturing, what equipment failure leads to the most blowouts?

38% of fracturing blowouts occur due to casing failure.

24% of frac blowouts occur when iron fails during frac flowback.

14% of frac blowouts occur due to wellhead failures.

7% of frac blowouts occur due to frac stack failures.

"An engineer's not going to put his hands on a fire, but he thinks he's so much smarter than us, and if they ever get a computer to cap a goddang oil well, I guess I'll be out of business. But I ain't shakin' in my boots over it."

Coots Matthews, Oilfield Firefighter[102]

Workover blowouts include loss of well control during the following:

- Pulling sucker rods
- Pulling/installing tubing
- Drill-outs / Clean-outs
- Routine pump change
- Changing out Electrical Submersible Pump (ESP) equipment
- Recompletion operations
- Snubbing
- Casing failure
- Swabbing

TACTICAL TIP

Never assume a well on artificial lift cannot flow naturally. Many wells on rods or ESP will flow up casing or tubing. It is a common mistake to let your guard down when working on a supposedly 'dead' well – assumed to be dead because it's on artificial lift, only to be unpleasantly surprised during the workover when you lose well control.

Blowout preventer (BOP) related incidents include problems with the BOPs that lead to a blowout, or problems when the BOPs are removed. Many operations teams think the risk of a blowout is insignificant once the production casing is thought to be successfully cemented. They let their guard down and turn attention to other activities, including preparations for mobilization to the next location. As soon as the BOPs are removed, Mother Nature has a nice surprise waiting in the form of a nasty well control situation. The situation can unfold similar to this:

1. Cement job is pumped with full returns
2. Floats hold
3. Cement is allowed to set
4. Well is assumed to be dead

5. Crew removes BOPs
6. Spraying fluid hits crew
7. Crew tries to stab BOPs back
8. Flow is too strong
9. Crew evacuates location
10. Blowout

TACTICAL TIP

Before rigging down BOPs, check for pressure and flow on the casing and annulus.

Although drilling blowouts account for the highest percentage of total blowouts from 1990 to 2015, over the past 5 years, fracturing-related blowouts have taken the top spot, accounting for the highest percentage of total blowouts, with 29%. Well control situations are often not thought to be associated with fracturing. However, the numbers suggest otherwise. Over the past 5 years, 70% of fracturing-related blowouts occurred as a result of casing failure (40%) or loss of control during flowback (30%).

Blowout Risk Statistics 2011 - 2015

Rank	Blowout During	Percent of Total Blowouts
1	Fracturing or Flowback	29%
2	Workover	19%
3	Drilling	19%
4	Production	17%
5	BOP Removal or Incident	9%
6	Wireline Operation	4%
7	Plug & Abandon	3%
8	Vehicle Collision	0%

"Every experienced drilling man can recite examples of foolish acts committed under the stress of potential blowouts by experienced personnel. Frequent practice in detecting simulated kicks and of closing procedures make correct moves a reflex action . . . . Most kicks that result in blowouts could have been controlled had the kick been discovered sooner, and the ease of killing the kick always depends on rapid discovery and well closure."

W.C. "Dub" Goins, Blowout Prevention Pioneer[103]

Blowout Preventers And Reaction Speed

Staying vigilant in this business is critical, especially when it comes to addressing well control situations. For example, you must be quick in reacting to a kick. When discussing the Macondo incident with others, primarily executive-type people, I frequently hear them say, *"The BOPs failed and it wouldn't have happened if they didn't."* This is flawed thinking because BOPs are not blowout stoppers, they are exactly what the name says – blowout preventers.

The topic of reaction speed was addressed at length during the Macondo trial. Oil and Gas Trial Lawyer representing Halliburton, Don Godwin, stated, "The Transocean Well Control Handbook, Your Honor, at the chapter dealing with actions taken upon taking a kick, containment as early as possible, says: 'When a well kicks, it should be shut in within the shortest possible time. By taking action quickly, the amount of formation fluid that enters the wellbore and the amount of drilling fluid expelled from the annulus is minimized. It is the driller's or the person performing the driller's function responsibility to shut in the well as quickly as possible if a kick is detected.' "[104]

"Next, Your Honor, the American Petroleum Institute, API, has a recommended practice for well control operations. Under the paragraph dealing with well close-in proceedings, it says, in part: 'When a kick is detected, the well should be closed in as quickly as possible to minimize influx volume.' "[105]

Expert Driller, Q&A:[106]

Q: "If you did suspect a kick, as the driller, you would shut in the well as quickly as possible?"

A: "Me myself, I would – if I suspected a kick, I would shut the well in. That's what I would – I would shut the well in."

"You don't fear your job, you respect it. You know your equipment and your men and you don't trust anybody but yourself."

Red Adair, Oilfield Firefighter[107]

Sr. Toolpusher, Q&A:[108]

Q: "Prior to 2010, was there ever a time on the Deepwater Horizon that it was required by the operations to close the BOP?"

A: "Any time we have an indication that something is not right, that's our first course of action."

Q: "Right. And that could be sensing of gas, acknowledgment of pressures. It could be many things, correct?"

A: "It could just be the driller's opinion that something is wrong."

Q: "Yeah. Many times, a perceived pending kick or pending blowout or pending loss of control of the well, you would shut the well in instinctively, correct?"

A: "The best way for me to put it is: A driller is trained to shut the well in if he has any indication that something is wrong."

Subsea Supervisor, Q&A:[109]

Q: "In the event that a well control event is detected, is there a protocol for informing you of that so that you can go to station to assist with shutting in the well?"

A: "Shutting in the well, secure the well, call me."

Q: "Okay."

A: "Don't wait on me."

Q: "Call you but don't wait on you?"

A: "No. Shut the well in."

Q: "Shut the well and then call you?"

A: "Then call me."

Drilling Expert, Q&A:[110]

Q: "Let's go back some steps. Our hypothetical driller has not yet shut in the well, but he has detected an anomaly that could indicate a kick. I'd like to ask your opinion and whether you believe his first step should be to try to discover what the anomaly is and then shut in the well, if it turns out it's a potential kick;

or is it your opinion that he should shut in the well and then try to study and determine what the anomaly is."

A: "Shut in – shut in first and then investigate. You don't try to investigate before you shut in."

Q: "Is that a basic well control principle?"

A: "Absolutely. It is taught in every school that I'm familiar with."

Q: "Is it – I'm asking this a different way. Is it a fundamental principle that every driller should understand? That is, if you have an anomaly, shut in the well, then determine what the anomaly is, as opposed to doing your study of the anomaly and then shutting in the well?"

A: "Once you detect an anomaly, you shut in the well and then investigate. And it's not the other way around."

Oil and Gas Lawyer representing Cameron states:[111]

"Transocean rigs had 409 kicks [from year 2005 to 2010]. And the average kick volume before the well was brought under control was 10 bbls. Transocean policy - and it's a good policy - states that the failure to limit the kick to less than 20 bbls is less than ideal. 89% of its kicks were caught within this 20 bbl limit.

Kicks over 20 bbls are called Code Red because they are critical events. Transocean had 60 of those. All but five, excluding Macondo, were under 100 bbls, and the largest was 200 bbls.

What was the size of the kick on April 20, 2010, on the Deepwater Horizon? Your Honor, it was literally off the charts. Depending on whether you use BP's estimate or whether you use Transocean's estimate, the Macondo well flowed between 700 and 1,000 bbls before there was even an attempt to close the BOP."

"When Transocean went before the national commission, the representative - its representative testified that the event was like a 550-ton freight train hitting the rig floor, followed by a jet engine's worth of gas coming out of the rotary. . . . But he also went on to say: 'It's like snipping a fire hose with a pair of scissors. The blind shear rams were not designed for that.' We certainly agree with that. Then he goes

on to say: 'And that jet engine of gas would have eroded away the rubber seals,' the packers I mentioned to Your Honor earlier, 'on the blind shear rams.' "

"In BP's accident investigation report, they discussed the fluid velocities that were ripping through the BOP when it was finally activated and was trying to close the well. It states that 'According to OLGA well flow modeling, the fluid velocity through a leaking annular preventer could have reached levels that were orders of magnitude greater than drill pipe steel erosion velocity.' In other words, Your Honor, velocity that could actually cut steel.

We believe the evidence will show that by the time the BOP was activated under these severe and extreme conditions - namely, a jet engine of gas moving at 10 times beyond steel erosion velocity - it was simply too late for this BOP, or any BOP for that matter, to be effective. So that eventually when the blind shear rams were in fact activated, the extreme conditions in the well ripped apart the BOP packers, making the BOP unable to seal the well."

Operations Analysis

There are industry experts and evidence that disagree with these statements, including evidence indicating that when the blocks fell, forces from above exerted on the drill pipe, and/or forces from hydrocarbon flow from below, or a combination, possibly pushed the drill pipe off center inhibiting the BSR from doing its job. Additionally, there were other contributing factors which could fill an entire book. In my opinion, the biggest contributor was failure to identify a kick and close the BOPs when the influx of hydrocarbons into the well initially occurred: a failure of early kick detection and quick response. Even if the BOP would have isolated the well, hydrocarbons were already above the BOP in the riser. The explosion, fire, loss of life, and associated equipment damage would most likely still have occurred.

"If quick, I survive. If not quick, I am lost. This is death."

Sun Tzu, The Art of War[112]

III

OPERATIONS CHECKLISTS

Aviation, space exploration, and other highly complex, capital intensive operations utilize checklists to manage risk. The oil and gas business, one of the most capital intensive industries, has much to gain by employing operational checklists. I am a strong advocate for checklists as they help ensure smooth, predictable, and efficient operations. Not utilizing checklists leads to problems for the following reasons:

1) Oilfield operations frequently include numerous difficult to remember steps and items.
2) Many people with various levels of experience, opinions, habits, and preferences are required to work together during implementation.
3) Key information is often forgotten during handoffs between day and night crews. It is easy to miss a small but critical aspect of an operation, especially after working a full day when the handoff meeting occurs with your relief.
4) Humans are emotional creatures that do not function at the same level of alertness throughout the workday.
5) It is easy to get distracted and disoriented once operations start. In many cases you are working in a grey fog of information and uncertainty. Checklists help deal with the adversity.

"Anytime you catch something just barely, where if you hadn't caught it you'd be in terrible trouble, you're using a checklist, even if not consciously."

Charlie Munger, Billionaire Investor[1]

The next time you see an astronaut working outside during a spacewalk, look for the small written checklist. Often it is strapped directly to their wrist and referred to as a "Cuff Checklist," that's how important it is. When your life is on the line, in addition to a lot of money, you cannot afford to forget anything.

Below are two NASA checklists regarding sample collection, drilling, and coring on the moon. Notice that NASA utilizes shorthand and abbreviations to save room because these checklists are directly on the spacesuits or in little books attached to the suit.

Apollo 16 Lunar Surface Cuff Checklist – Drill Core Sample[2]

1. Insert bit string in drill
2. Drill 3 sections into surface (steady drill, don't push)
3. Last section 8" off surface. Run drill 15 sec without further penetration

Apollo 16 Lunar Surface Cuff Checklist – Recover Core Sample[3]

1. Pull up drill as high as possible
2. Remove drill, cap top
3. Remove core by hand or jack
4. Cap bit
5. Core to Lunar Roving Vehicle, break between 3rd & 4th stem
6. Cap ends
7. Core stem to rack

Returning to the oilfield and Earth exploration, on the next page is a Cuff Checklist for Horizontal Shale completions. This checklist has been helpful to ensure key items are not overlooked prior to commencing fracturing.

"A NASA astronaut and a Russian cosmonaut can't be creative. He has to follow a predetermined detailed checklist written by an engineer and if he gets a little creative he'll never fly again."

Burt Rutan, Spacecraft Pioneer[4]

Horizontal Completion Cuff Checklist

Note: Actual checklist utilized in the field only includes the bold numbered headline words, everything else included below and check marked is for the readers' benefit.

1. **Plat**
   - ✓ Proposed well location survey
   - ✓ Surface hole, landing point, drill path, bottom hole
   - ✓ Section corner coordinates, directions, surface map
   - ✓ Proposed pad (all wells shown) and lease road easement
2. **Survey**
   - ✓ Directional drilling survey report (final version)
   - ✓ PDF and excel file types
3. **Plot**
   - ✓ Horizontal wellbore path map diagram
   - ✓ Hardline intersections noted
4. **Casing Tally**
   - ✓ PDF and excel file with details
5. **Mud Log**
   - ✓ Discuss with geology
6. **Gamma Ray**
   - ✓ Correlate to GR, find marker joint, compare to casing tally
7. **Offset Wells**
   - ✓ Map with all offset well locations identified
   - ✓ Notify offset operators
8. **Frac Procedure**
   - ✓ Send design and all required documents to team
9. **Frac Notice**
   - ✓ Send to state regulatory agency
10. **AFL (Lift and/or Tubing install) Procedure**
    - ✓ Send design and all required documents to team

To help provide a foundation for your own checklist construction, a start to finish "Lease to Grease Checklist" for drilling, completing, and producing one horizontal shale well within a U.S. commercial basin is presented on the next several pages.

Lease to Grease Checklist

1. Select risk tolerance: identify basin and formation of interest
    a. Frontier – no commercial production
    b. Commercial – current commercial oil and/or gas production within basin and from zone of interest
2. Perform basin wide analysis
    a. Build subsurface model
    b. Acquire seismic data
    c. Evaluate historic engineering and operations
    d. Perform production and economic analysis
    e. Assess surface topography
3. Perform political, legal, and economic risk assessment
    a. Assess oil and gas regulatory environment
    b. Identify all regulatory hurdles and economic impact
    c. Determine if state, county, and township is favorable for initial capital investment and full scale development
4. Rank acreage across basin for target interval
5. Determine strategy and interest by phase window
    a. Desired asset exposure (gas, condensate, oil)
6. Determine risk tolerance and identify area of interest
    a. Proved undeveloped (PUD) location – at least 90% certainty of commercial extraction
    b. Unproved location development
        i. Probable Reserves location – at least 50% certainty of commercial extraction
        ii. Possible Reserves location – at least 10% certainty of commercial extraction
7. Select prospective acreage position for deep dive analysis
8. Review well data on and offset to position
    a. Analyze production history
        i. Choke program and drawdown strategy

    - ii. Lift methods employed
    - iii. Estimate recovery
    - iv. Construct type curves
    - v. Run economics
  - b. Examine horizontal landing depths within target
    - i. Determine opportunity for improvement
  - c. Historic ability to maintain lateral within target window
    - i. Geosteering quality
    - ii. Re-steer wells, determine % within target
  - d. Investigate effective lateral length
    - i. Determine opportunity for longer or shorter laterals versus economic return
  - e. Analyze completion design
    - i. Determine opportunity for economic improvement with larger or smaller completion designs
    - ii. Estimate uplift potential
  - f. Understand historic costs and cost volatility
    - i. Determine opportunity for cost reduction
9. Model multiple economic scenarios
  - a. Flex initial production, decline curve shape, EUR, timing, commodity price, CAPEX, and OPEX
10. Construct detailed subsurface maps incorporating seismic
  - a. Cross section
  - b. Net pay isopach
  - c. Structure map
11. Identify geologic hazards and opportunities
  - a. Reservoir quality and deliverability
  - b. Faults and geologic anomalies
  - c. Water bearing zones (including potential sources for use)
  - d. Pore pressure and bottomhole static temperature (BHST)

"I do not seek, I find."

Pablo Picasso, Painter[5]

e. Shallow gas presence
f. High risk depths
    i. Overpressured intervals
    ii. Lost circulation zones
    iii. Difficult drilling depths: low ROP and bit life
    iv. Challenging hydraulic fracture initiation depths
    v. Corrosive intervals
    vi. Overburden and/or target interval instability during drilling, completion, and/or production
g. Knowledge of expected target formation thickness, target window, dip (updip or downdip), and target undulations
h. Expected oil quality and gas composition: NGLs, CO_2, H_2S, and/or rare gases

12. Craft water source and management strategy
    a. Determine optimal source for drilling and completion ops
    b. Sufficient and economic for full scale development
    c. Test water source for compatible
    d. Identify water reusability opportunities
13. Identify existing wellbore hazards
    a. Offset producing wells
    b. Offset disposal wells
    c. Offset inactive wells
    d. Offset cement plugged and abandoned wells
    e. Offset old mud plugged wells
14. Determine existing wellbore impact on offset development
    a. Pressure depletion
        i. Impact on IP and Type Curve shape
        ii. Difficult drilling (lost returns)
    b. EUR impact

"Every minute spent in planning and creating checklists will save you 10 minutes in execution and getting the job done."

Brian Tracy, Time Management[6]

c. Frac hits (positive or negative)
d. Well interference with development
    i. Ability to drill horizontal
    ii. Ability to frac lateral
e. Offset well hazard assessment
    i. Check all offset wells surface casing for pressure
    ii. Determine integrity of offset wells

15. Confirm risk and economics are attractive
    a. Perform Monte Carlo analysis
16. Check for acreage availability
    a. Off the ground leasing
    b. Farm-in / farm-out opportunity
    c. Asset acquisition
17. Perform thorough land-related due diligence
    a. Run title, ensure ownership of mineral rights for target formation and any secondary target formation potential
    b. Confirm no issues with obtaining appropriate surface use rights for development plan
    c. Evaluate forced pooling opportunities
18. Determine price willing to pay for acreage
19. Determine optimal development plan
20. Construct Geo Prognosis for 1st horizontal well
    a. Review logging program, coring program (if necessary), casing program, planned top of cement for each casing string, and wellhead design with team, including:
        i. Geologist
        ii. Drilling Engineer
        iii. Completions Engineer
        iv. Production Engineer
        v. Reservoir Engineer
    b. Run casing stress models to account for stimulation and production requirements

21. Engineer wellbore design and drilling procedure
    a. Determine depth to fresh water and approved surface casing setting depth
    b. Investigate possibility of shallow gas
22. Determine DFIT intervals
23. Run models to determine:
    a. Optimum fracture stimulation design
        i. Frac half-length
        ii. Fracture conductivity
        iii. Cluster spacing
    b. Tubing size optimization
24. Design artificial lift to be installed at the end of completion or later in the life of the well
25. Design surface facilities
    a. Research offset designs
    b. Model and optimize facility in simulator, working with vendors prior to construction
26. Estimate pipeline cost
27. Obtain consensus on future pricing assumptions
28. Evaluate takeaway capacity
29. Estimate total cost to drill, complete, produce and operate
    a. Run economics
30. If prospect passes economic hurdle, continue
    a. Lease, farm-in, and/or bid and acquire acreage
31. Acquire ample oil and gas operating insurance coverage
32. Determine list of vendors for all operations
    a. Bid work out to multiple vendors
    b. Perform due diligence on vendor service quality
        i. Visit with vendor personnel that will be on your location, discuss expectations and concerns
        ii. Contact references
        iii. Request to see vendor in action on competitor well

        iv. Acquire and evaluate designs, technical suggestions, and equipment information from vendor
        v. Research environment and safety records
    c. Negotiate favorable pricing and terms

33. Select DRILLING vendors for the following:
    a. Location Survey Services
    b. Rock Supplier – for location foundation
    c. Location Construction Services
    d. Drilling Consultants and Company Men
    e. Conductor Install - cellar, conductor, mouse, and rat hole
    f. Fuel Services – for drilling rig and other equipment
    g. Rig Mobilization Services
    h. Drilling Rig
    i. Rotating Head
    j. Drill Bits, Reamer, and Stabilizer
    k. Drilling Fluids
    l. Fluid Tanks
    m. Mud Gas Separator
    n. H_2S Safety Equipment
    o. Water Transport and Transfer
    p. Directional Drilling Services – MWD and Geosteering
    q. Mud Logging
    r. Rotary Steerable System – including insurance
    s. Electronic Drilling Recorder
    t. Drill Pipe and Drill Collars
    u. Casing Supplier
    v. Casing Inspection Services
    w. Casing Running Services
    x. Laydown Machine
    y. Centralizers
    z. Float Equipment
    aa. Toe Sleeve

bb. Cement Services
cc. Surface Casing Head
dd. Casing Spools
ee. Drilling spools
ff. BOP and BOP Testing
gg. Welding Services
hh. Communications (Phone, Internet, TV)
ii. Forklift
jj. Porta Potties
kk. Trash Trailer
ll. Trucking and Miscellaneous Mobilization Services
mm. Soil Farming Services
nn. Location Remediation Services

34. Select COMPLETION vendors for the following:
    a. Wellhead Equipment and Installation
    b. Cement Bond Log, Gamma Ray, CCL
    c. Water Transfer
    d. Above Ground Water Storage
    e. Frac Tanks and Trucking
    f. Acid Tanks
    g. Spill Containment
    h. Flowback Equipment
    i. Pump Down Services
    j. Frac Stack, Accumulator, Greasing Services
    k. Fracturing Services
    l. Fuel and Delivery
    m. Proppant and Delivery
    n. Fracturing Chemicals
    o. Wireline and Crane
    p. Perforating Services

"Rise early, work hard, and strike oil."
J. Paul Getty, Billionaire Oilman[7]

q. Cast Iron Bridge Plug
r. Frac Plugs
s. Light Plants
t. Man Lift
u. Forklift
v. RV Trailer and Generator
w. Heaters / Cool Down Equipment
x. Coil Tubing Services
y. Nitrogen Services
z. CT BHA Equipment
aa. Chemicals and Mixing Plant
bb. H_2S Trailer
cc. Filter System
dd. Transfer Pumps
ee. Gas Busters
ff. Fluid Pumps
gg. Tubing and Packer
hh. Pulling Unit
ii. Production Tree
jj. Production Facility Equipment and Installation
    i. Containment and Walls
    ii. Oil Tanks, Water Tanks, Bullet Tanks
    iii. Sand Separator
    iv. Separator, Heater Treater, Free Water Knockout
    v. Vapor Recovery Unit
    vi. Pipes, Fittings, Valves, Pumps
    vii. Lightning Protection
    viii. Electrical Equipment and Installation
    ix. Flare, Scrubber, Ignition System
    x. Compressors
    xi. SCADA System

35. Sign Master Service Agreement (MSA), reviewed by legal department, for all vendors and contractors
36. Submit formal Authority for Expenditure (AFE) with all capital requirements
    a. Include updated economic analysis and expected production performance
    b. Confirm all hurdle rates are met
    c. Obtain required internal and external approvals
37. Obtain gas purchase contract
38. File all state and federal required documents
39. Obtain government approval on all required permits
40. Write operations procedures
41. Build location, drill, and complete well
42. Notify state and/or federal regulatory agencies throughout process. File appropriate documents
43. Produce well; when ready, hook into facility and pipeline
44. Obtain oil, gas, water samples and send to lab
45. Monitor production and compare to initial projections
46. Employ choke management strategy to maximize economics
47. Perform post operations analysis
    a. Discuss all safety events and opportunities to reduce risk
    b. Identify opportunities to increase speed and efficiency
48. Share learnings across company, with service providers, customers, consultants, and company men
49. Incorporate experience into future processes and procedures
50. Drill, Frac, Produce, Improve, Repeat

"Sooner or later (usually later) the project phase comes to an end, all of the equipment has been hooked-up and the development phase commences. In terms of career advice, this is as good a time as any for the reservoir engineer to transfer from one company to another, preferably a long way away, for now the truth is about to be revealed."

L.P. Dake, The Practice of Reservoir Engineering[8]

IV

CORE TACTICS: 11 - 20

We continue on our journey, building a foundation of oilfield survival knowledge and intelligence with *Core Tactics: 11 to 20.* This section gets more technical and operationally complex. However, to help get key points across, the information and lessons are straightforward, direct and to the point.

The Core Tactics in this section continue to be a mix between drilling, completions, and production. The information is not siloed by discipline because many lessons span across the oilfield and touch different aspects of this business. I am strongly against isolating information or people within one discipline, even if the industry prefers to specialize people. It is not recommended, from a career perspective, to pigeonhole yourself by only learning information in your focus area. It is valuable to become an expert in one area, but also know enough about other disciplines to be able to make a career change later, if necessary or desired.

Additionally, understanding a train wreck in one area of the business can help prevent an incident within another discipline. The knowledge you gain is transferable and potentially stronger as you understand the various root causes of undesirable oilfield situations. It is more valuable for you to understand rather than memorize Core Tactics and apply them blindly within a specific segment of the oilfield. A professional "Tier One" oilfield operator would never do that.

"As CEO, I surround myself with people who challenge conventional wisdom."

Martin Craighead, CEO Baker Hughes[1]

11. ALWAYS: Pay Attention to Detail

If you only learn one thing from this book, it will be attention to detail. You will learn the value of attention to detail from the mistakes of others. If you are a big picture type person, not interested in the details and discipline an oilfield operation demands, it is probably best you stay out of this business. Find another job because you will not last long or you will hurt others. The business of oil and risk is not for everyone.

Attention to detail is hard. It takes effort and brain power. Many people do not want to do it. They just want to show up and pick up a check. If everything goes exactly as planned, maybe you can get away with just showing up. However, if you have worked in the oilfield for any significant length of time, you know the universal law of real world operations:

NO OPERATION GOES EXACTLY AS PLANNED

Small or large, there is always something that needs to be addressed and adjusted. Complex operations, with numerous moving parts and people, require an intense focus at many levels. Small issues can become big problems should they go unaddressed.

If you think your operation is consistently the exception to the universal law of real world operations, you are not looking closely enough and probably don't have all the facts. Ignorance is no excuse. You have to actively look for issues, ask questions, refine and optimize your operation to consistently avoid trouble and train wrecks, among a number of other undesirable oilfield situations. Preventing train wrecks is possible by paying attention to detail. It is in the small details of an operation that problems begin to reveal themselves, if you know what to look for and are watching relentlessly.

"God is in the details."

Ludwig Mies van der Rohe
Architect[2]

Warning Just Before Tubing Is Ejected

The job was to install 2-3/8 inch tubing on a producing well. The well had been producing up the casing after a frac job. Based on production performance, it was determined that tubing should be installed. "The well was still filled with a water/mud/oil solution which caused the well to be in a 'killed' condition. . . . The packer was installed at a depth of 9,500 feet by wireline through a 'frack stack.' After the packer was installed, the frac stack was removed and a BOP was bolted onto the wellhead flange. The well was then shut-in [for the night] by closing the main casing valve and the blind rams.

The following morning the well was opened, and an unremarkable amount of gas pressure was released. . . . The operation of installing the tubing through the BOP had begun, and 22 sections of 30 foot long tubing was run. On the 23rd joint, the movement of the tubing string stopped. The reason for the stoppage was unknown and totally unexpected."[3]

A discussion occurred between the operator, the service companies, the Packer Rep, and the Company Man. The plan was to "raise the string a few inches, then try to lower it again. [This operation was performed]. It again stopped, and the operator was instructed to raise it a few inches further [and try again]. When he did this, the well began to blowout the liquid contents of the well and the tubing. . . . The tubing came out in a 600-foot continuous string followed by the packer. . . . This took about 5 seconds. . . . The gas continued to blowout for 2 to 3 minutes. . . . One of the service company representatives managed to reach the main casing valve and close it . . . the pipe fell over and around the rig as it came out. All the employees ran, but a large loop section of the pipe fell on one employee, killing him instantly. Three employees with broken bones and soft tissue damage were taken to local hospitals."[4]

The investigation determined, "The fact that the 23rd joint of tubing hit something to cause it to stop down-hole movement suggests that the packer somehow moved up in the casing and lodged at a depth of about 600 feet. This occurred during the period following its installation [overnight] and the beginning of the 2-3/8" tubing installation. The packer was dislodged by the force of the tubing string [when it hit the packer]. The packer then became a 'piston,' forcing the tubing up and out of the well.

SURVIVAL SKILLS

When tubing is ejected from a well, it can happen very quickly and you may not be able to escape on foot.

One option, that has saved lives, is to take cover under something strong. Diving under equipment or a truck close by is an option. You may only have 1 second to determine the best course of action to save your life.

The packer was located several hundred feet from the wellhead embedded in the soft soil. . . . Another well in the same area of this blowout had a similar packer installed, which was discovered to have moved up in the well similar to the one at the accident scene, except that it did not rise up in the well as far, only a few hundred feet. The movement of the packer was discovered in this other case by running a 'slickline' tagger down the well with a lead weight to 'tag' the packer."[5]

Attempting to Outrun Iron

The previous fatality report warrants a discussion on the topic of iron and the efforts by good people to make an honest attempt to outrun iron when iron decides to search for a victim. Looking at iron and handling iron makes it seem heavy and slow, which is true when iron is acting alone. However, iron rarely works alone, which is why it can be very deceptive. Iron enjoys the company of gravity and pressure. When teaming up with pressure, iron can be faster than Usain Bolt, the fastest man in the world.

For example, when pipe is being ejected from a well, it often happens very fast. There are YouTube videos showing pipe being ejected slow. However, this is the exception. It usually comes out so fast that no one has time to record it. Iron combined with pressure transforms iron into missiles or bullets. When discussing with other field operators what to do if pipe is being ejected from a well or iron is coming apart, I generally say that I would take cover on the ground or under a truck, if I were lucky enough to react fast and take cover. Most of the time there is no time to react as these things happen in less than one second. Therefore, running may not help. However, you must always do what you think is best for the situation at hand. In certain situations, when iron is coming for you, running may be the best course of action.

Iron Faster Than Man

During a routine frac plug drill-out with stick pipe operation, the workstring unexpectedly parted as a connection was being made. At first, the tubing came up slow. The older hands dived under a water hauler truck that was parked alongside the pulling unit. A new on-job young trainee made a run for it into an open field.

As the workstring was ejected, it came down on top of the water hauler truck. One of the hands underneath the truck was hit and broke his leg. The other two hands under the truck were unharmed. The water hauler in the cab of the truck, watching everything unfold, was also unharmed. He saw the hands dive under his truck, so he did not try to move it. Unfortunately, the young trainee who ran into the open field was hit by parts of the ejected workstring, on the head and neck, and died instantly.

"During my many years as a drilling engineer & field consultant, I realized that much of what I had learned in the books did not apply to actual field operations."

Byron Davenport, President Kick Control[6]

12. NEVER: D&C Before Addressing Hazards

Everything we do in life contains an element of risk, known and unknown hazards with different levels of danger. Due to the capital intensity of this business, hazards must be minimized, monitored, and controlled, or drilling and completion (D&C) costs will quickly spiral out of control, at the very least.

To control oilfield risk, we must first know, and want to know, what the hazards are for the operation at hand. The Oilfield Survival Guide presents a 3-step process with an easy to remember acronym, **FAA**, **F**ind, **A**nalyze, and **A**ddress drilling and completion hazards before executing operations.

Find the hazards means searching for them, relentlessly. **Analyze** the hazards means getting to know them in detail, before they get to know you in detail. **Address** the hazards means scenario planning and taking action to ensure they do not destroy you or your profitability. Below is a list of key high-level pre-operations risks commonly overlooked or not adequately addressed prior to initiating drilling and completions operations. Consider using this checklist prior to each operation to confirm nothing has been overlooked.

Pre-Ops Hazard Assistant

1. Legal Hazards
    a. Mineral rights
    b. Surface use agreement
    c. Master Service Agreements
    d. Insurance coverage
    e. Legal hardlines
2. Geologic Hazards
    a. Faults
    b. Water bearing zones
    c. Pressure and BHST

    d. Shallow gas
    e. High risk depths
    f. Target window specifications
    g. H_2S presence
3. Offset Wellbore Hazards
    a. Producing and disposal wells
    b. Inactive and/or plugged wells
4. Offset Operations Hazards
    a. Drilling operations
    b. Workover operations
    c. Frac jobs
5. Technology Hazards
    a. Suboptimal engineering and well design
    b. Misapplied technology
    c. Untested and/or unproven technology
    d. Formation compatibility with drilling mud, frac fluid, and completion fluid
    e. Hydrocarbon compatibility with frac fluid
    f. Holistic implementation experience in area & target zone
    g. Service quality and consistency
6. Timing Hazards
    a. Upcoming holidays or special events
    b. Company disruptions
        i. Bankruptcy, merger, takeover, asset sale
        ii. Executive visits, field tours
        iii. Reorganizations
7. Weather Hazards
    a. Seasonal changes
    b. Rapid temperature changes
    c. Freezing conditions
    d. Extreme weather events
    e. Heavy rains / snow

8. Human Capital Hazards
    a. Experience level for task and technology
        i. Service company personnel
        ii. Operating company personnel
        iii. Company man and leadership
    b. Personnel distractions
        i. Turnover and new hires
        ii. Disruptive or highly emotional events
            1. Marriage, divorce, illness
            2. Interpersonal problems on the team
            3. Problems at home
9. Economic Hazards
    a. Budget and strategic plan
    b. Cost predictability / uncertainty
    c. Implementation risk
    d. Production performance uncertainty
    e. Commodity price volatility /collapse
10. Procedure Hazards
    a. Procedures constructed with full team involvement
    b. Incorporate technical and operational key points
    c. Address historic problems to prevent recurrence
    d. Job Safety Analysis written for each specific task
    e. Procedures abide by all government regulations
    f. Procedures follow all Company policies and requirements
    g. Approved by field and office management
    h. Reviewed on location before implementation

"Nothing happens only once in our universe."

Robert Hazen, The Story of Earth[7]

13. KNOW: The Exact Location of the Hardlines

Drilling or perforating across legal hardlines is a common mistake that can be very costly and time consuming to fix, operationally or in the courts. Check and double check the exact location of the hardlines and the rules in your area. Accepted practices can change significantly depending on the area of operations. You can lose an entire wellbore to the competition or a mineral owner if you are not careful.

> ***TACTICAL TIP***
>
> If you drill across a hardline, you may not want to do a wet shoe cement job because the casing shoe may then be considered the toe production depth. You would have to isolate it with a CIBP.

In some areas, you can drill across a hardline at the heel or toe without issues but cannot perforate across it. The way people get in trouble is that the drilling team, knowingly or unknowingly, drills across a hardline at the heel or the toe of a lateral. The completions team does not double check perforation locations in regards to the hardlines and perforates across them. The well is fractured and produced. Months later someone finds the error. Now your life will be filled with misery, and everyone in the company will know what you did.

If you cannot fix the well, at a bare minimum, you will have to spend many hours, if not days, at the oil and gas governing body of your State, listening to how you messed up while attempting to explain the situation to all the lawyers and government officials. As bad as things may seem, there is a worse situation. You drill a well and forget to double check the hardlines. The lateral ends up completely across the hardlines. Now you are in a very serious situation. Hopefully someone catches the mistake before you frac and produce the well because it is usually easier to deal with before the well is completed.

"Everything is funny as long as it is happening to somebody else."

Will Rogers, Actor[8]

However, you still have a major problem. One solution is to try and buy the acreage in which you drilled the illegal wellbore. Maybe you can get it cheap? If the owners do not want to sell, they have the upper hand and you are in a position of weakness. People have lost entire wells like this. Turn on the fear when it comes to hardlines because it pays to be a little paranoid to keep yourself out of trouble.

The Legend of Krampus Oil Corporation

Krampus Oil Corporation did not care much about hardlines. The head landman, Lauless Von Mumbles, would send out emails regarding hardlines that would ramble on about politics and news events. Buried deep within the rambles would be key land-related information regarding upcoming wells.

Around the winter holidays, Mr. Mumbles sent out an email regarding a change in the hardlines for an upcoming horizontal shale well. The hardlines would be moved 500 feet east of the western section line due to an agreement reached with offset operator, German Partners. Krampus Oil Corporation's lead geologist and lead drilling engineer were on vacation at the time. Their relief did not pick up on the change in hardlines, even though they were cc'ed on the email correspondence.

The horizontal well, Lucky Goat #1H, was drilled with the entire lateral across the hardline. German Partners found out by talking with the workers on the rig. They immediately called Krampus CEO, Lord Krampus. The reason German Partners was upset and the reason the hardlines were moved was because German Partners had 15 old stripper oil wells that were in the same zone as the new horizontal well. German Partners was concerned that their old wells would get damaged by getting hit with the Lucky Goat #1H frac.

"If a man didn't make mistakes he'd own the world in a month. But if he didn't profit by his mistakes he wouldn't own a blessed thing."

Edwin Lefèvre, Writer[9]

During the discussion, Mr. Krampus proposed to buy the offset section and all of the old wells from German Partners. A price was agreed to. However, Krampus Oil Corporation was almost bankrupt and was not able to pay for the wells without a big win on Lucky Goat. When the frac job commenced, everything was going as planned until the old stripper wells started to get hit with the frac. Instead of shutting down the frac stage, Krampus kept on pumping.

After the job was finished, they put the Lucky Goat #1H well on production with modest results. However, when they turned on the old stripper wells, each one produced significant amounts of oil. One by one, Krampus removed the old pump jacks as the wells were free-flowing. Once cleaned up, each of the old stripper wells went from about 5 bopd to over 1,000 bopd. Krampus Oil Corporation hit the jackpot with over 15,000 bopd from the old wells.

Operations Analysis

Lack of attention to detail is often the root cause of hardline crossings. This incident also had elements of poor communication and timing hazards, as the operation occurred around the holidays. Key team members were on vacation.

Most situations, in which a lateral is drilled completely across the hardlines, are not resolved as quickly as Krampus was able to do. You can easily get into a nasty legal situation where you will have to shell out big bucks. Fortunately, Krampus was able to turn the mistake into an opportunity, and it paid off handsomely. Most people that drill an entire well across the hardlines do not have this opportunity.

"Nobody is interested in following a man who, with his eyes fixed on the ground,
spends his life looking for the purse that fortune should put in his path.
The one who finds something no matter what it might be,
even if his intention were not to search for it,
at least arouses our curiosity,
if not our admiration."

Pablo Picasso, Painter[10]

14. UNDERSTAND: Your Contractual Agreements

Most aspects of the oil and gas industry involve some type of contractual agreement. Whether taking an oil and gas lease, working with a customer, or hiring a service provider, a legal document is often the foundation of the arrangement. Often, the field operations team, asset team, or subsurface team are not involved with drafting the contract, but work within its boundaries.

Take the time to obtain, read, and understand all agreements that may have an impact on your area of operation. It is not uncommon for these documents to not make it onto location or into the hands of the folks that need to understand them most. Whether you work on the services side of the business or on the operating side, in order to understand the contractual agreements that are in place, you will have to ask for them and invest time to understand them.

Do not assume someone will deliver contractual agreements to you, especially if an agreement has been in place for a long time. People forget the details in these documents, and over time, assumptions turn into unverified "facts" that become an unpleasant surprise when uncovered during a dispute or legal situation, as the following story illustrates.

Acquisition Team Fumbles on Lease Rights

Old Timer Offshore Corporation wanted to get into the shale game. They were evaluating an acquisition candidate in a basin that had stacked shale pay potential. The area of interest had two shale targets, both prospective. However, the lower target, Shale A, was the primary focus. Shale B was considered upside potential. The asking price from Shales "R" Unlimited, LLC was significantly above the Old Timer Offshore's valuation on the primary shale target.

"Those who tell the stories rule society."

Plato, Philiosopher[11]

In order to meet the acquisition price, Old Timer had to include the secondary target, but only 25% of the full scale development value. Old Timer still felt like they were getting a good deal. During discussions, Old Timer asked about the right to uphole potential and was told that it was included. The company performed some additional due diligence but not at an extreme level of detail.

Shales "R" Unlimited did not assign any value to Shale B. They justified the higher valuation based on a more aggressive commodity price outlook. Old Timer performed a significant amount of subsurface analysis on Shale B and thought it would be one of the most valuable shales in the world. As a result, they purchased the asset.

For two years, they focused on Shale A which delivered production performance as expected. Then they decided to start drilling Shale B wells. Initial rates were 3X the original valuation type curve rates. Old Timer thought they hit it big. However, what the land department soon uncovered was that they only held the right to drill Shale B on 50% of their acreage. The Shale B core area was held primarily by a competitor and nemesis of Old Timer. The Board of Directors was furious. The CEO of Old Timer, who was 89 years old, walked into an executive meeting on the issue and cracked his cane over the shoulder of the Vice President of A&D for the mistake.

Operations Analysis

Not understanding the lease contracts on the acquired asset cost Old Timer hundreds of millions of dollars. Poor due diligence and incorrect assumptions throughout the deal analysis and valuation process did not catch key facts. Never assume anything regarding contracts in this business. Due to the complexity and capital intensity of the oil and gas industry, the associated contracts are often just as complex. It is not uncommon to be surprised as to what language, terms, and conditions make their way into an oil and gas agreement.

"Trust, but verify."

Ronald Reagan, U.S. President[12]

15. STUDY: Offset Well History and Competitors

Before you drill, complete, recomplete, workover, or perform any type of operation, read the offset well histories. Learn from the experience of others before you start your operation. A lot of people never take the time to study offset wells or competitors. In my opinion, this is because they do not view the situation as owners. They are spending someone else's money, so why take the time to study?

Also, many folks in this business are over-confident, or even arrogant, and think they know everything. Instead of studying competitors and learning from them, they belittle and talk down about them. It is good to have confidence in your skills, but not to the point that competitors are completely dismissed. There is far too much money and risk on the line to not study the history of an area prior to performing any operations. Arrogance is expensive and dangerous in this business. It seems that the Earth likes to find these people and humble them, as the following story illustrates.

Identical Blowout Replicated by Offset Operator

A new shale play was discovered beneath a field with conventional production. To test the new shale play, operators were reentering the old vertical wells and drilling down into the new shale. In the overpressured (0.9 psi/ft) core area, Blaze Brothers, LLC reentered an old vertical well and drilled into the shale. They took a kick and shut the well in. However, the casing broke down at around 1,500 feet TVD. A nasty shallow underground blowout occurred, which took a month and $20,000,000 million dollars to control. A postmortem revealed that the old casing was corroded and could not handle the shut-in pressure from the kick.

A competing company, Loud'n'Proud Corporation, employed a fast follower strategy, realizing Blaze Brothers discovered the core. Loud'n'Proud leased up all the acreage around the blowout and prepared to drill their first test well, Inferno #1RE. They decided to re-enter one of the old vertical wells. During the meetings leading up to spud date, the ops team was making fun of Blaze Brothers blowout and how stupid they were.

Drilling Engineer: *"That would never happen to us."*

Drilling Foreman: *"They have no experience."*

Company Man: *"I used to work with their Company Man, and he is okay but not at our level."*

The geologist was worried and asked the drilling team if they reviewed the well file and job log.

Geologist: *"We have an interest in the Blaze Brothers' well, and their geologist sent me all the data. Lots of excellent information. You, guys, should study the files."*

Driller: *"Listen, little geologist, we know what we're doing. Blaze Brothers are idiots. Mind your business and don't worry about it. Just tell me where to put the rig and stay out of our way."*

Geologist: *"Okay, jackass." (muttered under his breath)*

Shortly after this meeting, Loud'n'Proud moved in and started drilling. In a week, they reached the shale. About 100 feet into the target interval, they took a kick which went unnoticed. By the time the crew realized what was happening, the rig was enveloped in a thick cloud of natural gas which quickly caught fire. They shut the shear rams, but the casing failed similarly to the Blaze Brothers blowout. However, what was different was that the gas came up backside, outside of the surface casing to the surface under the rig, encasing the rig in a sea of flames. The Driller barely escaped with his life. He lost most of his hair (it burned off), as well as his pride.

"Never give a sword to a man who can't dance."
Confucius, Teacher[13]

16. NEVER: Trap Yourself

From the moment you leave the house to go to work each day, there are many traps that must be avoided. On the way to location, it is easy to trap yourself and get into an accident. There are numerous ways a person can trap themselves on the road, an entire book could be written on it.

There are also many traps on location and ways you can unintentionally trap yourself just by walking across the pad. Positioning yourself between a backing vehicle and an object could be a death trap, don't do it. Additionally, there are unassuming things on location that look nonthreatening but could be a death trap. For example, innocent-looking mud pits can be a trap, depending on the situation, as the next two cases illustrate.

Death In A Pit of Burning Oil

"The crew had drilled approximately 2,000 feet and were preparing the hole to have it cemented. The cementing would be done to seal off the water layer underground. . . . While circulating water through the well prior to cementing, crude oil came out of the hole and ran off into the mud pit. . . . This left a trail of flammable crude oil from the rig down into the mud pit. Witnesses stated that the sediment pit contained approximately 4 feet of water with a 2-inch layer of crude oil now on top of it.

During the cementing portion of the operation, witnesses claimed that the cement bulk truck unexpectedly started to idle very fast. The truck's operator ran to the cab and shut off the engine. However, at the same time a small explosion occurred under the truck, and a wall of fire went from the truck to the wellhead. . . . The resulting explosion created a rupture in the engine area that allowed vapors under the engine compartment to be ignited. Once the rig was on fire, the oil trail leading from the rig to the pit caught on fire.

Prior to the fire occurring, a Drilling Employee was using a shovel to clean out the shallow ditch leading from the wellbore to the sediment pit. When the explosion and resulting fire occurred, the Drilling Employee was evacuating the area when he accidentally fell into the sediment pit. The pit did not have any guard or barrier around it. The distance from the pit's edge to the pipe trailer that was parked alongside of it was 10 feet. The distance was reduced by several feet because piping that had been recently used was laying on the ground next to the trailer.

Witnesses also said that the pit's liner extended several feet onto the ground, past the edge of the pit. Once the Drilling Employee fell into the pit, he was unable to escape it before the pit was consumed in fire, due to the flammable crude that was in it. The Drilling Employee's body was recovered several hours after the fire was extinguished and the pit was drained."[14]

Operations Analysis

Working around live oil always poses the threat of combustion. Volatile vapor coming off of the oil can easily ignite under the right conditions. In this particular case, once the fire started, the Drilling Employee tried to escape. A choke point, due to equipment between the trailer and pit in the chosen escape path, probably forced the Employee to run close to or on top of the pit's liner which extended out of the pit onto the ground and was most likely slippery. There was also no barrier around the pit. This combination of factors created a death trap. One key lesson from this incident is that you never know when you may have to escape from your current position. In certain situations, before working in an area on location, you may need to make adjustments, install equipment, or move things around so as to not trap yourself in case you need to get out of a situation quickly.

Fatality In The Foam

"A five man crew were drilling a new gas well and were at 354 feet. Drilling soap and water were used to lubricate the bit and to keep the dust down from the cuttings. The drilling soap and water normally create a certain small amount of foam when used. While drilling, [subsurface] water was encountered, creating a large amount of foam that was carried out the flowline and discharged into the containment pit. The large buildup of foam was noticed by the driller who sent two employees to the containment pit to beat the foam down before it came out of the pit.

The containment pit was approximately 80 feet long, 20 feet wide, and 14 feet deep. The containment pit was lined with a one-piece liner that was staked down at the edges of the pit. The pit contained an estimated 8 to 10 feet of water and 4 to 6 feet of foam. Prior to the accident, a fence was put at the front of the pit where the flowline discharges into it. The Foreman stated that the fence was put up to keep employees and anyone else away from the pit. The fence was put up from the cutout of the hill past the front of the pit and stopped at the down slope of the hill. Employee #1 and Employee #2 walked through the gap between the fence and the cutout of the hill and then walked back to the far left corner of the pit. The pit was not guarded along the sides or back where Employee #1 or Employee #2 walked or had stopped.

While both employees were at the far left corner of the pit, Employee #1 noticed that a 6 to 8 foot section of the liner, on the right side of the pit, had slipped down. Employee #1 walked around to the edge of the pit where the liner slipped down and attempted to pull it up. His feet slipped out from under him and he fell into the pit. Employee #2 ran around to where Employee #1 fell in. Employee #2 could not see Employee #1 due to the amount of foam and stuck his arm down in the pit to try and retrieve him. Employee #2 got a hold of Employee #1 a couple of times but could not hang on to him or pull him up. Employee #2 motioned to other employees that he needed help.

The Driller came over to the pit while another employee went to call 911. Employee #2 went to get a rope and returned. The Driller tied the rope around his waist and went down into the pit while Employee #2 held the rope. The Driller stated that he was waist deep in the fluid and the foam was over the top of his head. The Driller found Employee #1 with no movement. Employee #2 and another employee tried to pull up Employee #1 and the Driller, but the Driller could not hold into Employee #1. While down in the foam, the Driller stated that he was having problems breathing when the foam got in his lungs, mouth, and eyes, and that he was coughing and vomiting. The 911 rescue team arrived and after several attempts, Employee #1 was found and removed from the pit. Employee #1 was unresponsive and his body was taken off of the hill to the ambulance. The paramedics continued to try and revive Employee #1 but got no response. While being transported to the hospital by the paramedics, Employee #1 was pronounced dead."[15]

Operations Analysis

Recognizing a trap on location can be difficult. Visually, it may seem that foam within a mud pit is non-threatening. However, if you fall in, the combination is deadly, as this case illustrates. A pit's liner can make getting out nearly impossible because it's too slippery to grip. The foam can make it nearly impossible to breathe. Fall in, and you're fighting against both to survive.

"The lion cannot protect himself from traps, and the fox cannot defend himself from wolves. One must therefore be a fox to recognize traps, and a lion to frighten wolves."

Niccolò Machiavelli, Diplomat[16]

Death traps on location are often created through a combination of circumstances that can change during an operation. One minute, the trap does not exist; the next minute, it's there. Throughout an operation, assess each task from the perspective of how you could trap yourself while doing your job. Then take the necessary steps to manage the risk.

Backing Truck Traps And Kills Swamper

Let's review a vehicle incident, where a swamper was caught between a backing truck and the drawworks during rig down. Key excerpts (abridged and adjusted) from the fatality report are below:

"The crew arrived on site at 6:45 a.m. to begin rig down. . . . Mr. DeLorean was a swamper and was assisting the rig down. Swampers help chain loads down and hook the wench. . . . Mr. Heavyfoot was the driver assigned to move the drawworks. Mr. DeLorean went to the top of the ramp to guide Mr. Heavyfoot back. When the truck was in place, Mr. DeLorean would then attach the bridle to the hook, and Mr. Heavyfoot would operate the wench to pull the drawworks into the bed.

As Mr. Heavyfoot was backing up, the tires lost traction and [the truck] would not go back further [due to tires spinning]. Mr. Heavyfoot gunned the gas, the tires caught, and the truck crushed Mr. DeLorean between the back of the truck and the bottom of the drawworks. . . . Mr. DeLorean was taken to the hospital and was conscious and talking so they all thought it was not serious enough to result in a fatality. . . . Mr. DeLorean was in stable condition at the hospital but later that day died from injuries."[17]

Operations Analysis

Vehicle-related traps are often the most prevalent on location. Walking across the pad, the last thing on your mind is getting run over, but it happens, even by your own vehicle as the next case, *"Trapped and*

Killed By His Own Truck," demonstrates. Furthermore, positioning yourself between a backing vehicle and an object is incredibly dangerous. There are numerous things that can go wrong, resulting in getting crushed. Consider the possibilities listed below.

If the driver backing/parking the vehicle:

1) Does not see you (you are in a blind spot or fall down).
2) Misjudges the distance.
3) Presses the accelerator too much or foot slips and hits the gas.
4) Messes up and presses the gas instead of the brake pedal.
5) While parking, puts vehicle in wrong gear.
6) Misinterprets hand signals from the spotter.
7) Cannot hear voice commands from the spotter.
8) Sun gets in the eyes and cannot see for one second.
9) Fatigue or bad weather impacts ability.
10) Gets distracted (phone, radio, job, something on location, etc.)

If any of these things happen, and you place yourself in a precarious position on location, you can get trapped and crushed to death.

Trapped and Killed By His Own Truck

Let's review another vehicle-related fatality. Key excerpts (abridged and adjusted) from the fatality report are below:

"Todd exited his work truck to instruct a backhoe operator. He did not put the pickup truck in park. The truck started rolling. As he saw the truck moving, he tried to stop the truck by leaning against it. Todd was pushed back into the backhoe bucket and pinned between the truck and the backhoe, sustaining crushing injuries. Emergency medical personnel were called. They arrived and airlifted him to a hospital. At the hospital, he succumbed to his injuries and died."[18]

"Man is the only kind of varmint who sets his own trap, baits it, then steps in it."

John Steinbeck, Writer[19]

17. TWO: Tested Barriers

Redundancy is the key to longevity in this business. Having backup systems significantly minimizes the risk of a catastrophic event. Some folks will say that if a certain piece of equipment rarely fails, there is no point in having a backup. It may be true that a certain piece of equipment rarely fails. However, if it does fail, it could cause a mass disaster. Barrier equipment for pressure control is one of those critical areas in which equipment failure will result in disaster.

When working on wells, it is a good practice to always have two tested barriers between yourself and the reservoir. Below is a list of common activities to consider employing the two barrier rule:

- Maintenance on wellhead, frac stack, or production tree
- Removal of BOP, frac stack, or production tree
- Temporary isolation for simultaneous operations
- Tubing installation
- Recompletion or workover operations

A short list of barriers that must be tested prior to being considered as a barrier include:

- Cement
- Kill Weight Fluid
- Blowout Preventer
- Surface Valves
- Bridge Plug
- Packer with Knockout or Pump Out Plug
- Tubing Hanger w/2-way Check Valve

"One person's 'paranoia' is another person's engineering redundancy."

Marcus J. Ranum, Firewall Pioneer[20]

Cement Pumped, BOP Removed, Well Goes Boom

It was just another day in the oilfield. On this particular day, Hippie Energy Partners was focused on rigging down various equipment on the Lava Lamp #6H well for mobilization to the next location, while simultaneously waiting on production casing cement to set. The crew was not continuously monitoring the internal casing pressure or backside pressure because they were shorthanded and distracted with rig demobilization activities.

Additionally, there was another distraction. The charismatic, long-haired hippie CEO, Mr. Moonbeam Firepower, decided to visit location and congratulate the crew for a record spud-to-TD time. The backstory on how a long-haired hippie became the CEO of an oil company is legendary. Moonbeam Firepower was raising organic chickens on 20,000 acres in Texas, when the Eagle Ford was discovered on his land. Instead of leasing to an oil company, Mr. Firepower decided to get into the oil game. Over time, he became a wealthy oilman worth over $1.0 billion, as his acreage was in the core condensate window. Mr. Firepower was a bit lax on the safety side of the business. *"It's all in Mother Earth's hands. She blessed me with an ocean of oil!"*, he would say with a smile at every safety meeting, then walk off. That was the extent of his contribution to the safety meeting, and the hands loved it. They enjoyed working for him.

On this particular day, the crew was especially distracted by Moonbeam Firepower because he was handing out $1,000 gift certificates from the local gun store to everyone on location for the record spud-to-TD time. It is ironic that a hippie was essentially handing out guns for a job well done, but this was a Texas hippie, and Texas hippies love guns as much as they love bellbottoms.

At this point, the crew was sufficiently distracted and it was time to remove the BOP. While nippling down the BOP stack, it was discovered that the well had pressure on it. The crew quickly tried to stab the BOP back on the wellhead but were overcome with gas which quickly caught fire. The crew had no option but to run. Hippie Energy did not have a sign-in sheet, so they did not know how many people were on location. Since they were in the middle of a rig move, the location was crowded. When everyone heard Moonbeam was on site, workers from other locations also came to visit. There were 74 people on location when the blowout occurred.

Moonbeam ran into the fire to confirm no one was left behind. He rescued two employees and then went back to the rig to continue his search. Moonbeam pulled another worker from the fire, who had collapsed from the smoke, and brought him to safety. Moonbeam ran into the blowout for a third time. Unfortunately, he never returned. Moonbeam Firepower burned to death while searching for survivors.

Operations Analysis

With an accurate sign-in sheet, Moonbeam would not have gone back into the fire for a third time because he would have known that everyone was safe. Regarding the blowout, it is believed gas entered the wellbore through the shoe. When they bumped the cement pump-down plug, they bled back more than the typical 2 to 3 bbls (typical for their wells in this area). They bled back 15 bbls, and then the well flowed a small trickle of water. It was unimpressive flow and was dismissed. The well was shut-in. Gas built-up in the wellbore from the time they bumped the plug to the time they attempted to remove the BOP. They effectively had zero tested barriers. Two tested barriers would have prevented this tragedy.

"It is not the critic who counts; not the man who points out how the strong man stumbles, or where the doer of deeds could have done them better. The credit belongs to the man who is actually in the arena, whose face is marred by dust and sweat and blood...who comes short again and again, because there is no effort without error and shortcoming...if he fails, at least fails while daring greatly, so that his place shall never be with those cold and timid souls who neither know victory nor defeat."

Theodore Roosevelt, U.S. President[21]

18. SEE: It = Own It

If you see something unsafe, you are accountable. You own it. The situation becomes your responsibility. Do not walk past a problem and go about your day as though you saw nothing. It's not acceptable. Even if you have nothing to do with the situation, if you see something unsafe, you are obligated to address it and make it safe. It is understandable that you don't want to get into someone else's business on location. However, the business of safety is everyone's business.

Do not take this the wrong way, I am not suggesting to recklessly jump into the middle of an operation because you think it's unsafe. You could cause an accident if you wildly intervene. There is always the risk of unintended consequences. That must be taken into account on the more complex situations.

The question then is, when do you take action and get involved? When the little voice in your head says, *"That is dangerous, someone could get injured. I should say something. I should do something."* If you hear that inner voice, it's time to get involved. It's time to take action. Consider the following two scenarios.

Bit Salesman Slips On Rig Stairs Breaking Neck

The Toolpusher was leaving the doghouse, heading towards the stairs, when he slipped and almost fell down the stairs that were 23 feet off the ground. Fortunately, he managed to catch himself by grabbing the handrails. After calming down and gathering his thoughts, he noticed the metal walkway right before the stairs had a thin layer of grease, making it very slippery. He ran his boot across the metal, confirming how slippery it was. The Toolpusher, always in a rush, continued down the stairs.

"A person may cause evil to others not only by his actions but by his inaction, and in either case he is justly accountable to them for the injury."
John Stuart Mill, Thinker[22]

Thirty minutes later, a Drill Bit Salesman was leaving the rig floor and headed down the same flight of stairs. He also slipped on the greasy walkway right before the stairs. Unfortunately, he could not catch himself. He tumbled all the way down the stairs, breaking his neck. He was rushed to the hospital. His family was expecting him at his son's Little League game that afternoon. He never showed up. He died that evening in the operating room.

Toolpusher Cleans Up Grease, Preventing Fatality

The Toolpusher was leaving the doghouse, heading towards the stairs, when he slipped and almost fell down the stairs. He noticed the metal walkway right before the stairs had a thin layer of grease, making it slippery. The Toolpusher, although in a rush, decided to grab a few rags and wipe off the grease. He then called a quick meeting, explaining the near-miss to the rig crew. The crew took it upon themselves to clean all surfaces leading up to stairs, around the drilling rig.

Thirty minutes later, a Drill Bit Salesman was leaving the rig floor and headed down the same flight of stairs. He made it down the stairs without issue. He got in his truck and drove to his son's Little League game. His wife and daughter met up with them. They all went out to dinner that evening, never knowing what could have happened if the Toolpusher did not take ownership of a safety hazard.

Operations Analysis

Since the Toolpusher took action when he saw something unsafe, a fatality was prevented. Leading by example, he cleaned up the grease, demonstrating to the crew the importance of a safe work environment. Additionally, he shared his near-miss story, reinforcing the lesson.

"Unless the supervisors and their management evidence concern,
the crews upon whom the critical actions depend will also
become less concerned; and this state is the one
in which trouble is most likely to occur."

W.C. "Dub" Goins, Blowout Prevention Pioneer[23]

19. PERFORM: Due Diligence on All Partners

Operating companies and service companies are partners in implementing safe and effective operations. Operating companies are expected to provide a safe working environment, with safe procedures and processes. Service companies are expected to provide safe execution and high quality regarding the materials, equipment, and services. Both partners are expected to put safety first. Both must monitor each other's work and challenge each other if there is an issue.

Regarding the partner relationship on the aspect of service quality, do not assume the quality of equipment, materials, and lab testing is equal across the industry. Unless you have personally visited the local field maintenance yard, the workshop, and the lab you are using, you do not know the level of quality you are receiving. Before using a new vendor or technology, perform thorough due diligence.

New Vendor / Technology Train Wreck Avoidance Checklist

1) Meet the service provider team that will perform the work.
2) Ask to see them in action, if possible, on location or in the lab.
3) Audit the lab, workshop, yard, and equipment.
4) Identify where equipment and products are manufactured.
5) Obtain data on manufacturing defects and quality.
6) Check the company and crew safety record.
7) Obtain five customer references. Call each of them for insight on technology / vendor. Identify common themes. Craft opinion.
8) Ask vendor about train wrecks and the last few problem jobs.
9) Talk to the customers that lived through problems with the vendor / technology. How did they perform when things were not going as planned? What was done to solve problems?
10) Have vendor provide detailed quote / bid and obtain data showing how actual costs compare to estimated costs.

"Diligence is the mother of good fortune."

Miguel de Cervantes Saavedra, Military Leader[24]

Substandard Lab Found During Audit

Surprise Energy was having production difficulties related to fluid freezing (due to subzero temperatures) and surface oil emulsification. Speaking with chemical vendor, Black Magic, the decision was made to send samples to the main lab because the mobile lab on location did not have the proper equipment to perform the necessary testing.

There was only one problem, the main lab also did not have the proper equipment. In fact, the main lab did not have any equipment. A couple months prior, several scientists who worked in the main lab, left the company. Since all of the equipment belonged to them, when they quit, they took all of the equipment. Surprise Energy engineer, Gus Hoo, dropped by the lab unannounced to see how the tests were going, and found the lab completely empty of equipment. The few lab technicians, who were still working at the lab, were performing the testing at home, in their personal refrigerator and freezer.

Operations Analysis

If you are going to perform any type of lab testing, you must visit the lab to ensure it is up to par. During your visit, ask questions and get to know the scientists who will be working on your testing. If possible, visit when they are actually working on your samples. You can learn a lot by talking to the people in the lab.

These recommendations apply to any service. For example, when selecting a wellhead and frac stack vendor, go to their shop and look around. Ask the guys questions on the equipment. Look to see if the shop is well organized and maintained. Ask them to walk through their equipment inspection and quality assurance process. Depending on how things go, you may determine that the equipment and / or personnel are not optimal for your needs.

"Supposing is good, but finding out is better."

Mark Twain, Writer[25]

20. NEVER: Work on Pressurized Equipment

When working on any equipment, always confirm it has been de-energized. If something has the ability to contain pressure, always assume that there is pressure behind it. One area of the oilfield that has been responsible for a number of fatal accidents is working on iron that unknowingly has pressure behind it. Containing pressure is the job of iron. As sure as you are reading the OSG, iron is doing its job holding pressure. Check pressure gauges and bleed off pressure safely prior to initiating work. Gauges can fail, so having the ability to bleed off pressure can help confirm there is no pressure. If you need to work on a piece of equipment that could have pressure on it, check the gauges first, then see if it's possible to double check for pressure by opening a safe bleed-off valve.

Furthermore, if you check the gauges, then open a bleed-off valve, both indicating there is no pressure, then hammer on a piece of iron, or try to unscrew a connection, and have difficulty, you may still have pressure on the equipment. Stop what you are doing and check again. Pressure could be trapped somewhere in the system. For example, if there is an ice plug present, pressure could be behind the ice plug, and you would not see it on the gauge or when you open a bleed-off valve.

Flowline To Wellhead Installation Results In Fatality

Let's review a fatality related to the installation and connection of a flowline extension from a riser to a wellhead. Key excerpts (abridged and adjusted) from the fatality investigation are below:

"Workers were installing a pipe series onto the flowline to connect to the wellhead . . . the site had three separate wellheads approximately 5 feet apart. . . . Two employees of Solid Labor Roustabouts were connecting the piping. . . . The flowline riser was not straight, so the piping did not match up to the connection on the wellhead. Flex Wheeler, a manager of Match Energy Partners, told them to make the connection match up. Jack Handy, a manager of Natural Gas Connections heard the conversation and took it upon himself to use a handyman jack to lift the pipe connected to the flow stub. The flowline riser was pressurized to 2,100 psi and because of the brute force of the jack, the flowline fractured causing a catastrophic release, breaking off the piping . . . as the [4" pipeline] apparatus was blown off, it struck Mr. Handy who was thrown approximately 45 feet by pressurized gas, causing death by blunt force trauma."[26]

> ***TACTICAL TIP***
>
> Before touching any iron ask yourself if there is pressure behind this iron. Remember, the job of iron is to contain pressure.
>
> Therefore, until proven otherwise, assume iron is doing its job and containing pressure. Don't jack with it. Iron does not like to be messed with when doing its job.

Operations Analysis

Pressurized equipment is generally not designed to be worked on when it is containing pressure. Especially oilfield equipment, which is almost always not designed or engineered to be hammered on or moved when it is holding pressure. If you place a force on a piece of equipment that is under pressure, as Jack Handy did when he used the handyman jack to lift the pipe, there is a good chance it is going to blow up.

The accident investigation found that "there was not sufficient procedures and controls implemented to effectively ensure the pipeline was de-energized (flowline and riser blown down) before work began that led to a catastrophic release of natural gas."[27] Before touching any iron, if the crew would have checked for pressure, they would have seen that the flowline had 2,100 psi. That information would have been a big red flag not to touch anything. Until all pressure is released, bled to zero, and confirmed, no one should have begun work.

"The punch that knocks a man out is the punch that he doesn't see."

Cus D'Amato, Boxing Trainer[28]

Rig Walking System Kills Maintenance Man

"In the early morning hours, Employee #1 was called out to an oilfield drilling site, customer-operated, to perform troubleshooting and service repair on rig walking machinery. The employee was adjusting and replacing faulty parts, while the hydraulics system was under greater than 1,500 psi of pressure. The service union failed catastrophically without warning and blew apart, also launching a hand tool. These struck the employee in the face. The employee died hours later at the hospital. The hydraulics system had not been de-energized and isolated before the servicing."[29]

Removing BOP Flange Under Pressure Kills

"Two employees were directed by the Toolpusher, before the rig left the site, to remove the BOP flange from the wellhead so it could be used at another location. The flange was connected to the wellhead along with a 'Tulsa' ball valve to close in the well. As the crew were preparing to move the rig to another location, the two employees went into the cellar area . . . they had in their possession 2 chain wrenches . . . to hold the connection between the flange and valve assembly. As they began removing the flange by unscrewing the flange from the valve assembly, they inadvertently loosened the 7-inch piping below the Tulsa valve and nipple / union at the wellhead which was pressurized. The amount of pressure could not be determined during the investigation as there was no gauge installed on the wellhead.

> ***TACTICAL TIP***
>
> If you walk up to a wellhead on location, the 1st question you should ask yourself is, "How much pressure is on this well?"
>
> *Look for the pressure gauge. If there's no gauge, assume there's pressure on that well.*

While they continued to unscrew the flange, they were also unscrewing the piping from the wellhead until the last 2 remaining threads. At that time, the piping, Tulsa valve, and BOP flange all connected together, separated from the wellhead [as a result of natural gas pressure]. One worker was taken and hurled airborne 50 to 60 feet and fatally injured, while the other worker received serious lacerations and contusions."[30]

Operations Analysis

There are a number of lessons from the *"Removing BOP Flange Under Pressure Kills"* accident. First, the well should have had two tested barriers, not including the ball valve. Since the BOP flange was being removed and it was directly connected to the ball valve, the ball valve should not be considered a barrier in this situation.

Second, it seems they were removing a BOP flange component from the wellhead that had a threaded connection, and there was also a threaded connection below the ball valve. The accident report provides limited information on the wellhead configuration specifics. However, unscrewing one component above a valve, which has a connection below that is also threaded, would expose both connections to being loosened.

Lastly, it appears the crew assumed there was no pressure risk because there was no pressure gauge on the well to check for pressure. Pressure can build up anywhere it can get trapped, and often builds where and when you least expect it. Always have a pressure gauge to check before working on equipment. If you cannot check for pressure, do not perform the requested task, no matter who is asking you to do it. At the end of the day, it's not worth it.

"The danger which is least expected soonest comes to us."

Voltaire, Historian[31]

V

WRITING OILFIELD PROCEDURES

Written procedures are part of the foundation of a successful operation. Procedures must be tailored for each basin, target formation, and operations team performing the work. Additionally, procedures must abide by all federal and state laws, rules, and regulations. It is strongly discouraged to take a procedure from this book and blindly apply it to your area of operations. A professional "Tier One" operator would never do that.

In *OSG Volume One,* there are completions procedures for several operations as a reference or template. *OSG Volume Two* will include additional procedures from other disciplines. The following procedures are included in *Chapter 5*:

Completions Procedures

- Tubing and Packer Installation Procedure
- Diagnostic Fracture Injection Test (DFIT)
- Horizontal Plug & Perf Shale Completion Procedure
- Producing Well Temporary Isolation Procedure

These procedures are not perfect and are always a work in progress as specific issues come up. Additionally, procedures are always fine-tuned, adjusted, and rewritten during the crafting process as feedback is received from all team members. Below is a suggested 10-step process to consider when crafting your own procedures.

"Planning without action is futile, action without planning is fatal."

Cornelius Fichtner, Project Manager[1]

Building An Oilfield Procedure

1) Discuss upcoming operation with field personnel.
2) Discuss potential work with key service providers.
3) Bid work and select vendors.
4) Write preliminary procedure working with field consultants, company men, field superintendents, office engineers, and geologists.
5) Organize team meeting with service providers, field staff, and engineering to discuss preliminary procedure and build additional key details into procedure.
6) Update procedure based on the collective wisdom from the team meeting and send to management for feedback.
7) Send updated procedure to field consultants, field superintendents, geology, engineering, and all service providers for final comments.
8) Finalize procedure and obtain signatures from field management, office management, and Ops Vice President.
9) Execute operation. Track all positive and negative aspects of the written procedure relative to field implementation.
10) Update written procedure incorporating feedback and lessons learned from the actual operation to improve procedure for future operations.

"After a half century in the oil and gas business, I've learned a lot of lessons. Few have been cheap. . . .

A real leader never leads by fear.
I develop a rapport so no one is afraid to question my opinions or decisions.

They do it openly but respectfully, and I welcome it.

The best way to avoid confusion, misconceptions, and disasters is to have that fearless, open discussion.

If someone is afraid of me or afraid of making a mistake, they might be unwilling to speak up and provide critical information.

You can't afford that, especially in a high-stakes business like ours."

T. Boone Pickens, Billionaire Oilman[2]

As you can see, field operations and service providers should be involved in procedure construction. If the company man and vendors on location are not involved in crafting the procedure, you will definitely be missing a great opportunity to reduce risk and improve the probability of economic success. Do not write an oilfield procedure locked in your office, with no field involvement. Your procedure will either be ignored, or you will have problems.

The last two steps within *Building An Oilfield Procedure* are often overlooked. Folks in this business often forget to evaluate the operation and incorporate lessons learned into future written procedures. It seems we are all too busy to look back. However, I strongly urge you to view your procedures as a work in progress to be updated based on feedback from the field.

Let's walk through the construction of an oilfield procedure using tubing installation on a high pressure well for our first example. To provide context, this is a horizontal shale well that has just been hydraulically fractured, utilizing the plug and perf method, with composite frac plugs. We join the operation after coil tubing has just drilled out all of the plugs. The superintendent is a new hand and has requested a written procedure to install tubing on this high pressure well. We have been working with him, the company man, vendors, engineers and others to craft the procedure on the following pages. It has been signed off by Operations Vice President Jimbo H. Bondlog of Vesper Oil, so we are good to go. Read the procedure for yourself as I join you with reasoning and explanations in the knowledge boxes along the side.

"A man who does not think and plan long ahead will find trouble right at his door."

Confucius, Teacher[3]

Tubing And Packer Installation Procedure

Vesper Oil Corporation
J.W. Pepper 7-1H
Super Shale formation
AFE#1953

Jimbo H. Bondlog, Ops VP, reviewed and approved this procedure
December 2, 2016

Frank Underrig, Field Manager, reviewed and approved this procedure
December 5, 2016

> ***APPROVAL PROCESS***
>
> It's a good idea to list people that approved the procedure.

Objective: Install packer and tubing on the J.W. Pepper 7-1H well

Additional Attachments Include:

1) Wellsite Surface Schematic
2) Directional Survey
3) Casing Tally
4) Wellbore Diagram
5) Contingency Plans

> ***ATTACHMENTS***
>
> List of accompanying documents helps ensure nothing is overlooked. It is also a good idea to PDF everything to prevent someone from accidentally deleting something when handled on location.

Directions to Location:
From the intersection of Highway 495 and RM Highway in Deer Park, turn South and travel 5.0 miles down lease road to location on right.

Wellsite:
4 horizontal wells on location
J.W. Pepper 7-1H is the most northern well
(see wellsite surface schematic)

> ***CONTACT INFORMATION***
>
> Consider putting your name and phone number at the bottom of each page to encourage communication when you are not on location, or when you are sleeping and need to be woken up ☺

Page (1 of 7)

Engineer in Charge:	Daniel Bernoulli	Cell (405) 555-5555

Casing Detail:

OD (in)	Wt. Lbs/ft	Grade/ Connection	Depth Start (ft)	Depth End (ft)	Drift (in)	ID (in)	Cap. bbls/ft
13-3/8"	61#	J-55/BTC	25	1,500		12.5	0.1521
9-5/8"	47#	P-110/BTC	25	9,000		8.68	0.0732
5-1/2"	20#	P-110/Hyd	25	20,000	4.653	4.778	0.0221

Marker Joints located at 7,000' and 14,000' (see casing tally)

Additional Well Data:

TVD: 11,000 ft

MD: 20,000 ft

BHT: 260 °F

WELL DETAILS

Provide key well details for your operation. Always include bottomhole temperature (BHT). Many oilfield components have temperature limits and are sensitive at certain temperatures, especially when exceeding 250°F.

Safety Rules:

- Hold safety meetings every day prior to work
- Hold task specific safety meetings throughout the day as required
- Confirm all personnel are prepared and equipped to work
- Find unsafe situations, equipment, procedures and fix them
- Never work in unsafe conditions or do anything unsafe
- Remind all personnel of **Stop Work Authority** rights & obligations:

  1. **Anyone Can Shut The Job Down For Safety**
  2. **You Are Required To Shut The Job Down If Unsafe**

Vesper Oil Corporation encourages anyone that thinks the operation is unsafe to exercise **Stop Work Authority** and shut the job down immediately. You are not only encouraged to shut the job down if unsafe, but required to shut the job down. You must shut the job down immediately. We will only work when everything is safe. That is my commitment to you. – Linda Vesper, CEO Vesper Oil

Page (2 of 7)

Verify At Pre-Operations Safety Meetings:

1) Number of personnel on location
2) Percentage of "Short Service" employees is below limit
3) Proper PPE for all operations
4) Everyone has radios that needs them
5) Weather related issues are addressed
6) Emergency Plan of Action
7) Emergency drivers and vehicles identified
    a. Proper vehicle and informed driver
    b. Ensure full tank of fuel

Deer Park Hospital: 101 Robert Moses Blvd.: (405)-555-5555

Life Flight: 1-800-555-5555 (Air Rescue)

Packer and Tubing Installation Procedure:

For the readers benefit, fully decoded acronyms are included. Short hand, oilfield lingo, and area specific terminology have been minimized as much as possible.

1) Walk across entire location, clean surface, confirm safe worksite, and proper site preparation for operations. Utilize vacuum truck if necessary. Have several 24/7 vacuum truck company contacts, should operation require cleanup services on short notice.

> ***EMERGENCY INFO***
>
> Include your emergency numbers and emergency plan in the procedure. It's an opportunity to remind the team that safety is the most important part of the operation.

2) Identify all potential operational hazards and discuss during safety meeting and throughout the day.

3) Identify and Mark high risk 'Hot Zones'.

Page (3 of 7)

Engineer in Charge:	*Daniel Bernoulli*	*Cell (405)555-5555*

4) Confirm wellhead and frac stack are clean and in good working order with clear and clean cellar access.

> **TACTICAL TIP**
>
> A clean wellhead, stack, and BOP helps you see and identify leaks and other issues quickly before they get out of control.

5) Confirm wellbore is clean and prepared for Junk Basket / Gauge Ring run. Flush casing and/or perform additional Coil Tubing (CT) runs to guarantee vertical portion of well is free of sand, plug parts, or other debris to avoid problems during wireline operations.

> **RISK MANAGEMENT**
>
> A simple tubing installation can turn into a complex snubbing operation if you get stuck running in the hole and need to fish. A clean wellbore is cheap insurance.

6) Ensure wellbore is full of fluid and record wellhead pressure. Provide all details in daily report each morning.

7) Record dimensions and take photos of all items entering wellbore.

8) Move in and rig up (MIRU) wireline and crane. Ensure wireline & tools are well maintained, in good working order.

9) Rig Up (RU) Lubricator, and pressure test. Equalize lubricator pressure with wellhead pressure.

10) Run Gauge Ring and Junket Basket to 8,700' measured depth (MD).

11) Report tight spots and debris found in basket (take photo of debris found in basket, send to engineering). If issues identified, text/call ops engineer: Daniel Bernoulli

12) Construct the following assembly (bottom 'a' to top 'f'):
    a. Wireline re-entry guide with 3,000 psi (differential pressure) pump-out plug
    b. 2' pup joint
    c. XN-nipple
    d. 6' pup joint
    e. 2-3/8" x 5-1/2" retrievable production packer
    f. Stinger portion of On-Off tool w/X-profile

13) Confirm calculations on pump-out plug settings are correct.

14) Run in hole (RIH) with packer assembly to 8,500' MD and set packer mid casing joint.

15) Pull out of hole (POOH) with wireline.

16) Bleed pressure to 0 psi and monitor for 30 minutes to ensure well is dead. If pressure will not bleed off call ops engineer before proceeding.

17) Lubricate in tubing hanger with 2-way check valve.

1st BARRIER

Once the packer with pump out plug is successfully negative pressure tested, it becomes the first method of defense.

18) Ensure pressure is zero, monitor for 30 minutes to confirm.

19) Rig down wireline and release from location.

20) Nipple down (ND) frac stack. Leave closed lower master valve in place. Install night cap with needle valve and pressure gauges.

2nd BARRIER

Once the tubing hanger and 2-way check valve are installed and successfully negative pressure tested, it becomes the second method of defense. Always have two methods of protection.

Page (5 of 7)

21) After all wells on location are at this point, packers installed with two forms of well control, proceed with program.

22) MIRU workover (WO) unit.

23) Confirm pressure is 0 psi, then ND night cap.

24) NU BOP on lower master valve.

25) Record pressure test on BOP.

26) Lubricate out tubing hanger with 2-way check valve.

27) Rack and tally 8,600 ft of 2-3/8" 4.7# L-80 EUE tubing.

28) RIH with the following:
    a. Overshot portion of On/Off tool
    b. 1-joint of 2-3/8" 4.7# L-80
    c. X-nipple
    d. 8,500 ft of 2-3/8" 4.7# L-80 to surface

29) RIH with 2-3/8" tubing to 8,500 ft. Tag packer. Pickup 30 ft circulate packer fluid.

30) RIH 30 ft and latch into packer. Space out and land tubing with tubing hanger and 2-way check valve. Tighten lock-down screws.

31) Ensure pressure is 0 psi. Monitor for 30 minutes.

TACTICAL TIP

Don't overlook the proper packer fluid. It is often forgotten.

32) ND BOP and lower master valve. NU production tree.

Page (6 of 7)

33) Pressure test production tree against 2-way check valve. Increase pressure in 500 psi increments until 7,500 psi. Monitor for leaks during this process.

34) Lubricate out 2-way check valve.

35) Once all wells on pad are at this point, proceed to next steps.

36) Close production tree valves. Install pumping tee on top valve.

37) Pressure up on tubing and shear pump-out plug. Monitor annulus pressure and tubing pressure. Record in daily report.

38) Remove pumping tee, RD WO unit and release.

TACTICAL TIP

Don't forget to pressure test the production tree. This procedural step is commonly overlooked.

TACTICAL TIP

An oilfield professional must always be aware of what is happening on the backside. Always keep a good eye on any unusual indications. When you pump-out the plug, do not fixate on the tubing and ignore a potential warning sign on the annulus in the form of unexpected pressure.

39) Connect to production facilities and turn pad over to production. Report estimated first production date in daily report.

40) Safety first.

***** END OF PROCEDURE *****

(Page 7 of 7)

Engineer in Charge:	*Daniel Bernoulli*	*Cell (405)555-5555*

Diagnostic Fracture Injection Test Procedure

ShaleX Resources

Xtreme Shale #1H

AFE#2016

Mr. Apollo, Director Resource Development, reviewed and approved this procedure January 2, 2017

Objective: Test casing integrity for fracturing and perform DFIT on toe of horizontal exploration well.

Casing Detail:

OD (in)	Wt. Lbs/ft	Grade/ Connection	Depth Start (ft)	Depth End (ft)	Drift (in)	ID (in)	Cap. bbls/ft
9-5/8"	40#	J-55/LTC	25	1,500		8.835	
7"	29#	P-110/BTC	25	9,000		6.184	0.03710
4-1/2"	13.5#	P-110/LTC	25	15,000	3.795	3.920	0.01492

Additional Well Data:

TVD: 10,000 ft

MD: 15,000 ft

BHT: 225 °F

Tie Back Receptacle: 8,900 ft

Float Collar: 14,920 ft

1) Clean and dress location.
2) Check wellhead for pressure. If well has pressure call engineering before proceeding.
3) ND wellhead cap, NU tubing head and two 7-1/16" 10K frac valves.
4) MIRU wireline. Check for pressure and document readings. Pull Retrievable Bridge Plug (RBP). Run CBL from as deep as possible in the curve to 1,000' above the top of cement.
5) RD wireline. MIRU CT. (Page 1 of 4 - DFIT Procedure)

6) Confirm casing and annulus are filled with fluid.

7) Perform pressure test as follows:

8) First increase pressure on 4.5" x 7" annulus to 3,000 psi. Hold 3,000 psi on annulus.

9) While holding 3,000 psi on annulus, increase pressure on 4.5" production casing to 9,000 psi. Confirm pressures are stable for 10 minutes. Text picture of pressure test to 'Ops Group Text List.' If pressures are not holding call office.

10) Slowly bleed pressure on 4.5" production casing to 0 psi; then slowly bleed pressure on annulus to 0 psi. Monitor for 30 minutes. Confirm pressure is 0 psi on casing and annulus.

OPERATION ORDER

The configuration on this particular wellbore requires the annulus to be pressured up before pressure is placed on the casing. This is to prevent the tieback casing from being pumped out of the receptacle.

Each wellbore is different, and you must know the differential pressure you can place across components.

For certain things, the implementation order is just as important as the task.

11) MU CT BHA and RIH to clean out to FC. Tag FC and circulate wellbore clean. POOH

Suggested CT BHA: (Take pic of BHA on location)

a. 3.75" 4-blade concave junk mill
b. 3" XO sub
c. 2-7/8" motor
d. 2-7/8" agitator
e. 2-7/8" hydraulic disconnect
f. 2-7/8" dual flapper valve

(Page 2 of 4 – DFIT Procedure)

TECH DETAILS

If you want something rigged up in a certain way, include the technical details.

12) PU TCP guns with 2 ft of perforations (6 spf) 12 shots total 60° phasing.

13) Tag FC to correlate depth. Perforate 1 cluster from 14,890' to 14,892'. POOH. RD CT.

14) RU WL and run 15K bottom hole pressure gauges to 8,500' MD.

15) MIRU pump truck.

16) Install self-powered surface pressure monitoring gauge.

17) Pressure test surface lines to 6,000 psi. Set electronic kickout to 5,500 psi.

18) 5,200 psi is maximum pressure for DFIT.

TACTICAL TIP

Having surface and downhole gauges provides a fail-safe in case one system has problems. This puts you in the best position to have usable data.

19) Increase annulus pressure to 300 psi for DFIT.

20) Pump treated water at 3 bpm until break down occurs. Once breakdown is achieved, pump an additional 10 bbls of treated water and shut down for DIFT. Send pic text to "Ops Group Text List." Once pumps are shut down, do not start injecting again.

21) Obtain ISIP and 5, 10, 15 minute pressures.

22) RDMO pump truck. Send all data to engineering.

TACTICAL TIP

Although this Injectivity Test looks simple, it is very easy to mess it up and have inconclusive or unusable data as a result.

23) Clean location and install protection around well.

(Page 3 of 4 – DFIT Procedure)

24) Obtain data from surface gauge on daily basis for analysis to determine when gauges can be pulled. To determine reservoir properties in a shale well using DIFT's, the well needs to be shut in for at least one month, sometimes even longer.

25) Safety first, thank you.

***** END OF DFIT PROCEDURE *****

(Page 4 of 4 – DFIT Procedure)

PROCEDURE TIP

Always include the page number and total number of pages in the procedure.

Additionally, consider putting a sentence at the end of the procedure to indicate it is the last step and that the procedure is finished. Here, we included *"Safety first, thank you."* Another option is to include an end note *"End of DFIT Procedure."*

There have been a number of train wrecks due to losing a page of a procedure, the order of the pages getting put together incorrectly when printed in the field, or two procedures getting mixed together on location.

"Could I do it [tell someone how to find oil] in one sentence? . . .
The closest I can come to that sentence beyond
'Listen to the Earth'
is that you have to get down under & beyond
the mere occupational greed and
look into the simplicity, the purity,
the sacred part of it – the act,
not the results,
and yourself."
Rick Bass, Geologist[4]

Horizontal Plug And Perf Shale Completion Procedure

Millennium Resources
Tataouine CE 3PO-1H
Solar Shale, Orion S. System
AFE#14-R2D2-13B

Mr. Lucas Yodea, Ops VP, reviewed and approved this procedure May 25, 2017

Objective: 45 stage plug & perf frac stimulation on Tataouine 1H well

Additional Attachments Include:

1) Wellsite Surface Schematic
2) Directional Survey
3) Casing Tally
4) Wellbore Diagram
5) Casing Movements Simulation
6) Contingency Plans
7) Perforation Schedule
8) Pump Schedule
9) As-drilled Hardline Crossing Survey
10) Gamma Ray and Mud Log
11) Geologic analysis with fault identification
12) Offset wells analysis and monitoring plan

Directions to Location:

Within the Milky Way Galaxy, Orion Spur, 77 light years from the Outer Rim, travel to Port Skikda. Location is 62.5 miles south from Mos Zarzi. Coordinates: 33°50'34.1"N 7°46'44.7"E

Wellsite:

8 horizontal wells on location
Tataouine 1H is most southern well
(see wellsite surface schematic)

Casing Detail:

OD (in)	Wt. Lbs/ft	Grade/ Connection	Depth Start (ft)	Depth End (ft)	Drift (in)	ID (in)	Cap. bbls/ft
13-3/8"	61#	J-55/BTC	25	1,500		12.5	0.1521
9-5/8"	47#	P-110/BTC	25	9,000		8.68	0.0732
5-1/2"	20#	P-110/Hyd	25	20,000	4.653	4.778	0.0221

Marker Joints located at 7,000' and 14,000' (see casing tally)

80% Burst on 5.5" Casing = 10,100 psi

80% Burst on 9-5/8" Casing = 7,550 psi

Additional Well Data:

TVD: 11,000 ft

MD: 20,000 ft

BHT: 250 °F

DLS Rules of Thumb (when pumping down frac plugs):

10°- 15°: No Issues

16°- 20°: Dangerous

>20°: Severe Situation

HISTORIC PROBLEMS

Here, I list dogleg severity because this area has a history of problems when the DLS exceeds 15°.

Your experience from past issues in an area should be incorporated into your procedures.

Never assume people know or remember past oilfield problem jobs.

Hardline: This wellbore crosses perf hardline at 19,900' MD

Safety Points:

I. Safety is the Highest Priority.

II. Stop Work Authority is a right and obligation by all those who are working on this location.

III. Hold safety meetings each morning and as needed for individual operations.

IV. Sign-in and account for everyone. Always know how many people are on location.

V. Max Rate is **75 BPM.**

VI. Max Pressure is **10,000 psi.**

(Page 2 of 8)

VII. Set Electronic Kick-Outs staged: **9,700 to 9,900 psi**
VIII. Set Main-Line Pop-off (full open full bore) at **10,100 psi.**
IX. Monitor and record surface casing pressure.
X. Closest hospital is Mos Zarzi. Flight time is 10 minutes.

Completion Detail

Effective Lateral Length	# of Stages	Stage Interval	Lbs./ Stage	BBL/ Stage	Clusters/ Stage	Rate (bpm)	Fluid Type
9,000 ft	45	200 ft	300,000	10,000	8	75	S.W.

S.W. = Slickwater; Ensure 20# gel is available if needed

Procedure:

For the readers benefit, fully decoded acronyms are included. Short hand, oilfield lingo, and area specific terminology have been minimized as much as possible.

DESIGN SUMMARY

Including a high-level overview of the completion design can help the field team see the big picture before diving into the details.

1) Confirm you are rigged up on the correct well.
    a. Reference Wellsite Schematic
    b. Confirm with cellar welded well number nameplate
    c. Finalize with GR and location of marker joint. Each well on location has the marker joint at 200' MD difference from other wells and must be confirmed.
2) Install spill containment for tanks and completion equipment.
3) Set frac tanks and manifold together including 8" tees and valves. Obtain water samples and perform testing per policy.
4) Record pressure on all casing strings.
5) Move In Rig Up (MIRU) wireline and 15K test unit.
6) Run gauge ring and junk basket until tools will no longer fall. Report any tight spots.

7) Run in with GR/CCL/CBL until tools will no longer fall. Log until 1,000' above Top of Cement.
    a. Correlate to Gamma Ray (GR). Then find marker joint. Compare location to Casing Talley. If off more than 5 feet call engineering. Send bond log electronic file to engineering for analysis.

8) Rig up pump down crew. Pressure up to 7,000 psi in steps of 1,000 psi. Toe sleeve pinned to open at 7,000 psi. Do not exceed 9,000 psi.

> ***TACTICAL TIP***
>
> Confirming good cement is a critical step before the frac job.

9) Once sleeve opens, establish injection at 10 bpm. Once rate is established, pump 500 gallons of 10% HCl at max rate flushed with treated water.

10) Run in with Cast Iron Bridge Plug (CIBP) and perforation guns to 19,890' MD and set the CIBP.

11) Pressure test 5.5" X 9-5/8" annulus to 1,000 psi for 30 minutes. Monitor casing pressure.

> ***TOE FRAC OPTION***
>
> Instead of perforating stage 1, you could design the completion to frac out of the sleeves. If you choose this option, make sure you get a good max pressure test prior to fracturing.

12) With 1,000 psi on annulus. Pressure casing to 11,000 psi and hold pressure for 30 minutes. Monitor annulus pressures.

13) Once pressure test confirmed successful: stable pressure for 30 minutes; first bleed casing pressure off from 11,000 psi to 0 psi. Then bleed pressure off annulus.

14) Perforate Stage #1 according to "Perforation Schedule."

(Page 4 of 8)

15) Record pressure increase if any. Pull out of hole (POOH).

16) Lubricate in tubing hanger and back pressure valve. Confirm back pressure and tubing hanger are holding pressure (monitor for 30 minutes).

17) Close lower frac valve. The lower frac valve and tubing hanger and back pressure valve serve as two forms of isolation.

18) Nipple Up 15K frac tree as follows (from bottom to top). Utilize nipple up crew and hydraulic tools. Scribe bolts after torqueing. Label number of turns required on wheel valves.
    a. 7-1/16" 15K Full Opening, Manual Frac Valve (already in place)
    b. 7-1/16" 15K Full Opening Hydraulic Frac Valve (HCR)
    c. 7-1/16" 15K Cross with 4-1/16", 15K Wing Valves (HCR inside and manual outside)
    d. 7-1/16" 15K Full Opening Manual Frac Valve
    e. 15K Frac Head with Coil Tubing Top Entry Access

19) Ensure all wellhead ports, pins, and plugs are fully engaged and rated for 15K. Everything must be rated for 15K.

20) Rig Up 15K flowback manifold and bleed-off line per Company Policy.

FRAC STACK DETAILS

There are many ways to construct the frac stack. Each area and company has a preferred method.

For this area, 15K iron is preferred because treatment pressures are sometimes above 10K and the team prefers the added security and optionality. Additionally, for this area, due to historic problems with the HCR, they prefer to use the top "swab" valve as the primary valve.

(Page 5 of 8)

21) Install full open pop-off on 5-1/2" X 9-5/8" annulus set at 2,000 psi. During frac job maintain 1,000 psi on annulus. Monitor and record pressure during entire operation.

22) Remove tubing hanger and back pressure valve.

23) Rig up test unit and test frac tree and flowback equipment to 15,000 psi. Step pressure up in steps of 1,000 psi. Hold pressure test for 30 minutes.

24) Provide valve maintenance per manufacturers recommendations. (See attached valve greasing procedure).

25) Set up water transfer to move water at 75 bpm with 100% backup.

26) Witness blender bucket test to ensure proper delivery of chemical additives.

27) Frac stage 1 according to Pump Schedule.
    a. Apply 1,000 psi on annulus during frac.
    b. Max Pressure is 10,000 psi.
    c. Set Electronic Kick-Outs staged: 9,700 to 9,900 psi
    d. Set Main-Line Pop-off at 10,100 psi.
    e. Max Rate is 75 BPM.
    f. Flush to bottom perforation cluster.

Notes: Lower Manual Frac Valve will not be used during frac job. Do not use the Hydraulic Valve as a primary means of closing in the well during the frac operation for routine operations.

- Use the top "Swab" Valve as the Primary Working Valve.
- Hydraulic Valve is for Emergency situations.
- Ensure accumulator is out of danger zone.
- Do not perform hard shutdown at end of flush.
  (Page 6 of 8)

- On first 4 stages if >50% probability of screenout, pump a sweep stage and reassess.
- After 4 stages, the well should have adequate productivity to flowback screenouts.
- If screenout occurs, do not try to re-establish injection. Flowback a minimum of 1.5 wellbore volumes before trying to re-establish injection.
- Obtain ISIP and 5 minute pressure after each stage.

SWEEPS

Sweeps are a tool you can use to reduce screenout risk. Some companies do not like them. Always check with your boss regarding company preferences and strategies.

My view is that screenouts can start a chain reaction of events that cause undesirable oilfield situations to occur.

- Hold Plug Pump Down Meeting with team to establish excellent communication and process, when running in hole with plug and guns, to prevent premature setting of plug.

28) Repeat steps on remaining 44 stages (45 Total).

29) Utilize 12K lubricator (or greater). Test lubricator to 9,000 psi. Function test wireline BOP. Perforate each stage per Perforate Schedule.

30) Torque and test frac stack every 5 stages. Ensure grease is actually being injected and correct amount is being used.

31) Rig down stimulation equipment.

32) MIRU 2" Coil Unit. NU CT BOP's. Test BOP's and lubricator against lower frac valve to 8,000 psi.

33) Follow "Drill Out Procedure" for running in hole, drilling plugs, and running out of hole. **(Page 7 of 8)**

34) Preferred CT BHA:
    a. 3.75" Concave Junk Mill w/centralizer
    b. 3" XO Sub
    c. 2-7/8" Motor
    d. 2-7/8" Agitator
    e. 2-7/8" Hydraulic Disconnect
    f. 2-7/8" Dual Flapper Valve

35) Open well to flowback tanks.

36) When well begins to produce hydrocarbons, turn well through facilities. Place well on production and perform State Potential Test. Obtain oil and gas samples for analysis.

37) Safety first.

***** END OF PROCEDURE *****

(Page 8 of 8)

"Safety is part of our culture...It is not only a priority but also a core value."

Dave Lesar, CEO Halliburton[5]

Producing Well Temporary Isolation Procedure

The following procedure is to prepare the location for additional drilling activities. The objective is to:

A. Install and test two forms of subsurface isolation.
B. Remove above ground equipment.
C. Install a nightcap with pressure monitoring and bleed off capability.
D. Install fall protection and surface barriers.
E. Confirm location is ready for drilling activities.

> ***JOB OBJECTIVES***
>
> Listing primary objectives can help the field team understand the end goal of the operation.
>
> This helps when changes need to be made on location in real-time, during troubleshooting. With the end goal clearly listed, the field team can do what is necessary without losing sight of the objectives.

Procedure:

1) Clean and dress location. Test anchors.
2) Obtain pressure on tubing and annulus.
3) Move in rig up wireline and crane.
4) Rig up lubricator and pressure test to 5,000 psi.
5) Equalize lubricator pressure to well pressure.
6) RIH with wireline retrievable blanking plug to 10,000 ft and set in XN nipple. Pull out of hole.
7) Ensure well is full of fluid and confirm well is dead.

> ***TACTICAL TIP***
>
> Forgetting to equalize pressure is a common mistake that often results in a fishing operation.
>
> Train wrecks tend to have small beginnings that snowball into an avalanche of problems.

a. Bleed pressure to zero and monitor tubing, tubing annulus, and casing annulus for 30 minutes.

1st BARRIER

The tested blanking plug is the first barrier.

8) Nipple up lubricator and install 2-way check valve.

9) Rig down wireline.

10) Monitor well for 30 minutes to confirm well is dead.

2nd BARRIER

The tested 2-way check is the second barrier.

11) MIRU workover rig and associated equipment.

12) ND production tree.

13) Nipple up, test, and chart BOP.

14) Lubricate out 2-way check valve.

15) Unlatch off packer and trip out w/ 10,000' of 2-7/8" tubing.

16) Make up 5-1/2" Retrievable Bridge Plug (RBP) and setting tool. PU and install lubricator and RBP assembly. Run in RBP to 7,000'. Set RBP and POOH. Verify well is dead. RD lubricator. Release wireline. Fill well with water and biocide.

17) Monitor casing and annulus for 30 minutes to confirm well is dead.

18) Nipple down BOP and wellhead to below ground level. Install night cap with a pressure gauge and needle valve. RDMO rig.

19) Install cage, fall protection, and barricades around well.

20) Flush and blank off flowlines. Remove surface lines.

21) Clean location and prepare for drilling rig mobilization to pad.

***** END OF PROCEDURE *****

VI

CORE TACTICS: 21 - 30

In terms of technical difficulty, capital intensity, uncertainty, risk, and discovery, the business of Earth exploration, in the form of oil and gas operations, is rivaled only by the business of space exploration. From a certain point of view, space exploration is easier in comparison because space is empty; the Earth is not.

When we perform space exploration, what is the point of it? We do it because we, humans, are explorers. That fact will never change. And what's the first thing we do after space travel to another planet, be it the Moon or Mars? We collect rocks and DRILL. When naysayers proclaim that the oil and gas business is old, outdated, and will be replaced, smile and laugh at them.

When the Earth gets replaced, then this business will get replaced. Frac bans and other regulations may slow us down, but it means nothing in geologic time. The oil and gas will be produced sooner or later, on this planet and the next. Long after we all have turned to dust.

"Where were you when I laid the foundations of the Earth?
Tell me, if you know so much."
Book of Job[1]

Earth's Oil and Gas Timeline[2]

Event	Age (years ago)
Universe Formed	14 billion
Sun Formed	5.0 billion
Earth Formed	4.5 billion
Utica Shale Deposited	460 million
Marcellus Shale Deposited	380 million
Bakken and Woodford Shales	360 million
Mississippian Limestone	345 million
Barnett and Fayetteville Shales	325 million
Wolfcamp and Bone Spring	300 million
Hanifa Source Rocks Deposited	153 million
Haynesville and Bossier Shales	150 million
Shilaif Source Rocks Deposited	96 million
Eagle Ford and Austin Chalk	92 million
Monterey Shale	6 million
Neanderthals Utilize Oil	70,000
Oil is refined by Ali ibn al-Abbas	Yr-950
1st Shale Gas Well, New York	Yr-1821
Pennsylvania Oil Boom Begins	Yr-1860
Oklahoma's 1st Commercial Oil Well	Yr-1897
Texas Oil Boom Begins	Yr-1901
Bolivar is Discovered, Venezuela	Yr-1917
Gachsaran is Discovered, Iran	Yr-1928
Burgan is Discovered, Kuwait	Yr-1937
Ghawar is Discovered, Saudi Arabia	Yr-1948
Fyodorovskoye is Discovered, Russia	Yr-1971
C.W. Slay #1 Barnett Shale Discovery	Yr-1981
Shale Boom Begins, Barnett	Yr-2002
Marcellus Shale Renz #1 Discovery	Yr-2003
Fayetteville Shale Boom Begins	Yr-2004
Bakken Shale Boom Begins	Yr-2006
Haynesville Shale Boom Begins	Yr-2008
Eagle Ford Shale Boom Begins	Yr-2010
Utica Shale Boom Begins	Yr-2011
Shale Boom Goes Global, Vaca Muerta	Yr-2012
Woodford Shale Boom Begins	Yr-2013
Shale Boom Goes Bust	Yr-2015

21. ALWAYS: Have an Escape Plan

Thorough escape planning can help avoid situations in which you become trapped. Practice helps prevent panic under pressure when a well control incident, or similar situation, occurs. Consider taking the time each day to contemplate what you would do in an emergency situation. Play the scenario out in your mind. Scenario planning can help reduce the potential that you will freeze or do something unintended when things are happening in real life.

> ***FACT***
>
> 34% of American Workers Do Not Feel Prepared for an Emergency Situation
>
> *NSC Survey* [3]

Below is the 5-step OSG escape plan roadmap:

Emergency Escape Planning

1) Write out multiple scenarios on how a given oilfield operation could turn into an emergency situation.
    a. Short hand, just a few notes on what could happen.
2) For each scenario, consider what you would do to try and prevent the situation from spiraling out of control.
3) Accept that you may not be able to fix the problem.
4) Plan and practice the escape (construct a blueprint).
    a. How will you escape? Review your Company's plan.
    b. Are you prepared to run, if onshore?
    c. Are you prepared to jump (if necessary), if offshore?
    d. Where will you run /jump from and to?
    e. Discuss and coordinate with all personnel to prevent chaos and escape accidents.
    f. Play the escape scenario out in your mind.
    g. Practice a simulated escape. Run escape drills.

"You don't ever go in someplace before you figure out how the hell you're gonna get your ass back out."

Red Adair, Oilfield Firefighter[4]

5) Plan next steps: Once safely away from the danger, confirm everyone is safe. Get medical attention to all injured personnel.
    a. Ensure you have a current personnel list.
    b. Confirm you have correct location coordinates.
    c. Confirm emergency phone numbers.
    d. Confirm cell phone reception. If it's sporadic, find an area where cell reception is dependable.
    e. Determine how and where you will meet the ambulance or helicopter. Is there a safe place for the helicopter to land?
    f. Familiarize yourself with the location of the closest hospital and the time it will take to get there.
    g. Confirm hospital has a trauma center equipped to handle severe injuries.

Accepting that an operation may not go exactly as planned puts you in a position of strength to deal with adversity. Accepting that problems could occur is the first step; planning on what action to take is the second. Incorporating escape planning into your daily routine will substantially improve the probability that you and your team will go home to your families if an emergency situation occurs.

When operations go bad, there may not be time to conduct the escape plan while the emergency is happening. Escape planning needs to occur before the emergency. Most folks involved in a significant incident or accident, simply cannot believe that it happened to them or that they had witnessed the situation unfold. Complete disbelieve is commonly the first reaction of most folks involved in major train wrecks. Never think that any situation described in this book cannot happen to you.

"The first hour after injury [The Golden Hour] will largely determine a critically-injured person's chances for survival."

R Adams Cowley, Father of Trauma Medicine[5]

Panicked Pusher Kills Crew And Company Man

The job for the day was to run a packer and tubing on a recently fracture-stimulated horizontal well, Shenanigans #18-3H. George Gibbler, Chotskies Oil & Gas Corporation engineer, sent out the procedure to Company Man Willie Lumberg on Friday, then shut off his phone for the long Memorial Day weekend. The procedure only had one method of isolation for the operation, a burst disc in the packer.

Lumberg recognized the mistake, as Chotskies Corporation has a minimum 2 barrier policy. Chotskies management, as recently as one week prior to the accident, distributed buttons to all employees to wear that state "Love 2 Barriers" – to help employees remember and promote the two-barrier policy. Lumberg called Gibbler multiple times to discuss his concerns but was not able to get him on the phone. Lumberg then discussed the issue at length with Milton, Toolpusher with Swingline Energy Services. Milton thought that it was not a big deal because *"the odds that the packer will fail are almost zero percent,"* he stated to Lumberg. Additionally, Milton had plane tickets for a trip to Mexico and did not want to miss his flight due to a delay in obtaining a second barrier for the operation at hand.

Lumberg and Milton decided to proceed with the operation. The packer and tubing were run without issue. Next, the BOP's and lower master frac valve were removed to install the production tree. As soon as the lower frac valve was removed, the well started to flow. Initially, flow rate was only a trickle of water. This occurred for around 5 minutes. After 8 minutes, flow rate increased significantly. The decision was made to quickly reinstall the lower frac valve. As the crew attempted to isolate the well, an eruption occurred, blowing fluid and gas up over the top of the workover rig derrick.

"Chance favors only the prepared mind."

Louis Pasteur, Chemist[6]

In a panic, Milton the Toolpusher, ran to the closest available vehicle, a lifted pickup truck, to escape location as natural gas was quickly swirling around the well, the workover rig, and everyone on location. Milton slammed on the accelerator, speeding away from the well. However, in the process, he unintentionally ran over two Swingline employees and Lumberg, the Company Man.

Once everyone congregated at the upwind emergency meeting area, a headcount was performed by Robbie Bobby, packer hand. Four of seven people were accounted for. Robbie witnessed Milton run over three people during the chaotic escape. He informed Milton of his actions. Milton denied it and had no recollection of what happened. Robbie's story was collaborated by crane man, T.P. Stevenson.

Local emergency personnel arrived on location within 30 minutes; the blowout subsided in less than one hour, allowing for the retrieval of the three men who were pronounced dead due to multiple fatal injuries. The well was successfully isolated and work finished later in the week by a second crew. Based on injuries and further investigation by city police, all evidence indicates Milton killed the three oilfield workers during the chaotic escape as he was driving away from the uncontrolled fluid release, striking them with the heavy duty iron front bumper and running over them with the vehicle at a high rate of speed.

Operations Analysis

Two tested barriers would likely have prevented the accident. Also, if they would have taken the "trickle" flow seriously when initially noticed, the accident could have been avoided. There is no difference between a trickle and a full flowing well because both can become a blowout. Finally, if Milton would have prepared himself to deal with an emergency situation, he might not have panicked during the escape.

"Early in my career, I learned that the only predictable thing about this industry is its unpredictability. I also recognized that no one knows what the future holds, but those who are prepared for the future are the ones that succeed."

Jonny Jones, CEO Jones Energy[7]

22. ALWAYS: Inspect Your Pipe

Casing, drill pipe, tubing, and workstring problems are responsible for some of the most expensive train wrecks in the oilfield. Spending the money and taking the time to thoroughly inspect your pipe is money well spent. Never short-change or skip the inspection process. Additionally, perform the engineering and analysis to ensure you are utilizing the optimal tubular design for your operation.

Let's study a court case where alleged workstring imperfections, high concentrations of hydrogen sulfide, and operational decisions on location contribute to a blowout.

Alleged Pipe Imperfections Contribute to Accident

"The well was 'sour,' meaning that it contained a high concentration of hydrogen sulfide, and PowerSour Operating sought to complete the well while it was 'under pressure.' PowerSour retained Snubberlovers Services to provide snubbing services and a snubbing unit, which is a hydraulic workover rig used to push down or pull up on piping or other equipment inside the well. PowerSour rented from Manny Rentals a 'workstring,' which is used, in part, to operate equipment inside the well. The workstring was comprised of used joints of 'T-95' piping, which is a type of piping specified under the National Association of Corrosion Engineers (NACE) 'standard MR0175' as appropriate for hydrogen sulfide environments. This piping had a represented tensile strength of 171,200 psi.

After Snubberlovers had successfully set a packer . . . problems arose with releasing the setting tool from the packer. After some initial attempts to release the setting tool from the packer proved unsuccessful, well operations were shut down for the evening. PowerSour decided to flow the well during the overnight hours to release pressure in the well. Flowing the well necessarily involved releasing from inside the well a 'nitrogen blanket,' which was a protective layer of nitrogen that was pumped into the well to displace the hydrogen sulfide to protect surface workers in the event of an emergency and the piping from the corrosive effects of the H_2S.

The next morning, Snubberlovers, under the direction of PowerSour, used its snubbing unit to pull up the well piping in incrementally increasing amounts as part of the continued effort to release the setting tool from the packer. When Snubberlovers pulled the piping at between 128,000 and 130,000 pounds of pressure, the piping parted at 'Joint 18,' which was located approximately 500 feet below the surface. It is undisputed that the piping broke as a result of sulfide-stress cracking, a process that occurs in hydrogen sulfide environments which causes pipe to become brittle. After Joint 18 parted, the piping then sprung up the well, slamming the snubbing unit with over 1,000,000 pounds of kinetic force. The snubbing unit had an engineered failure point of 600,000 pounds, and the piping ejected from the well, causing property damage but no personal injuries.

PowerSour sued Manny Rentals and Snubberlovers for negligence, breach of contract, breach of warranty, and products liability. Snubberlovers and Manny Rentals generally denied PowerSour allegations and asserted claims against PowerSour for breach of contract and negligence.

At trial, the parties presented substantial, conflicting testimony regarding the cause of the sulfide-stress cracking in the piping and the ejection of piping from the well. PowerSour presented evidence that Manny Rentals furnished piping that was damaged from prior use, not of the quality represented, susceptible to sulfide-stress cracking, and unsuitable for the sour well. They also presented evidence that Manny Rentals was aware that the piping that it provided was to be exposed to a hydrogen sulfide environment and would not be continuously protected by a nitrogen blanket. For example, PowerSour Vice President of Drilling, Paul Pitting, testified that they needed piping that could 'meet certain criteria and withstand the environment' in the well,

Manny Rentals was aware of the well's sour environment, no one from Manny Rentals ever stated that a nitrogen blanket was necessary to protect the piping, and PowerSour would not have used the piping had it known of any use-restrictions. Mr. Pitting asserted that the only purpose for using a nitrogen blanket is to protect 'the safety of personnel,' and he opined that it was 'absurd' to believe that a nitrogen blanket is used to protect equipment and piping. Mr. Pitting also testified that the piping provided had some 'pretty significant' scarring from prior use and was defective because it failed at 75% of the represented tensile strength.

Mr. McCracken, PowerSour's metallurgical expert, testified that T-95 pipe is specifically tested to withstand a 'nasty hydrogen sulfide environment.' In reviewing photographs of the failed joint of piping, McCracken noted 'significant indentations' on the pipe's surface, and he explained that 'no tubular with such indentations should have been run into the well given the known downhole conditions.' McCracken noted that the well's environment would have caused sulfide-stress cracking on non-resistant materials, and he opined that the presence of the indentations on the piping 'basically turn[ed] that originally non-vulnerable material into a vulnerable material because of the cold working that occurred as a result of the indentations.'

PowerSour alleged that Snubberlovers was negligent during its snubbing operations and in using a snubbing unit that failed and allowed the broken pipe to eject from the well. PowerSour introduced evidence that Snubberlovers had participated in inspecting Manny Rentals piping before putting it into the well. Mr. Pitting testified that the Snubberlovers supervisor at the well, Mr. Gouges, had rejected a few of the joints for 'deep scarring' from prior jobs. Mr. Pitting noted that although Mr. Gouges' job reports, which were introduced into evidence, reflected that he had 'concerns for using Manny Rentals equipment' and did not feel 100% safe with Manny Rentals pipe, he had not shared these concerns with PowerSour. PowerSour also introduced evidence that Snubberlovers had not informed PowerSour that the snubbing unit had an engineered failure point that permitted certain parts of the unit to fail and, in this case, allowed pipe to eject from the well.

Through their evidence, Snubberlovers and Manny Rentals presented a causation theory that was directly opposed to that presented by PowerSour. Snubberlovers and Manny Rentals alleged that the joint on the workstring had failed as a result of PowerSour's critical error in deciding to remove the nitrogen blanket that had been protecting the piping. Although Manny Rentals expert, Dr. Yolanda Yield, testified that T-95 pipe is appropriate for use in a hydrogen sulfide well, she disagreed with PowerSour's contention that T-95 pipe will 'always work' in such an environment and would not be susceptible to sulfide-stress cracking. She noted that, under the governing industry standards, the user of the pipe, which in this case was PowerSour, should have also considered the concentration of the hydrogen sulfide and the 'partial pressure' in the well.

Yolanda Yield explained that NACE standard MR01075 provides that approved materials, like T-95 piping, are resistant to sulfide-stress cracking 'under defined conditions,' but are 'not necessarily immune to cracking, meaning there is still a possibility that they can fail.' She opined that the fact that the joint broke at a level of applied pressure that was below the pipe's rated tensile strength did not establish that the pipe was defective because tensile strength values are 'determined in the air,' not in hydrogen sulfide environments like that presented in the well. Thus, Yolanda Yield, through her testimony, indicated that T-95 pipe that is not otherwise defective can fail at rates lower than the represented tensile strength ratings depending on the environment surrounding the pipe.

Ms. Yield further explained that a well operator like PowerSour can avoid sulfide-stress cracking of piping in sour wells by controlling the surrounding environment. As examples of how to control the environment, he explained that an operator could use a chemical 'that mitigates the environment' or could 'separate the environment from the material.' Specifically, Yolanda Yield noted that an operator could use a nitrogen blanket to reduce or eliminate the concentration of hydrogen sulfide from the areas of a well that have 'the greatest susceptibility to cracking.' She explained that in determining the areas of piping that are most susceptible to breaking, one would need to consider factors like depth, temperature, and partial pressure. She noted that Joint 18, which was located in the top third of the well, was located in the area of the well that was most susceptible to sulfide-stress cracking.

Yolanda Yield opined that the decision to remove the nitrogen blanket from the well 'let the hydrogen sulfide contact the material and the area which has the very high stresses up high in the hole.' She further opined that if PowerSour had not removed the nitrogen blanket, the pipe would not have parted as a result of sulfide-stress cracking because the nitrogen would have excluded the H_2S.

Snubberlovers and Manny Rentals also presented evidence that the T-95 pipe, before being used in the well, had been inspected and declared to meet American Petroleum Institute (API) specifications. PowerSour's Paul Pitting also testified that, after the pipe had ejected from the well, PowerSour retained a metallurgist who tested the piping, and this testing revealed that it 'met the API.' Additionally, there is evidence that when Manny Rentals piping arrived at the well, it was visually inspected by PowerSour Company Man, Justin Doit, and PowerSour's own snubbing expert, Noll Problemo, along with Snubberlovers supervisor, Mr. Gouges. Mr. Gouges testified that when the piping arrived at the well, it was marked as premium API inspected piping. He noted that neither Justin Doit nor Noll Problemo had ever expressed any concerns that the piping had excessive scarring. In fact, Mr. Gouges noted that when he personally decided to 'kick out' a few of the joints for 'deep gouging,' Mr. Doit objected and telephoned Mr.

Bigg Wheel, a PowerSour' superintendent in Houston. Mr. Wheel, instructed Mr. Gouges to run all of the piping 'anyway' because it had been 'API inspected.' Despite Mr. Bigg Wheel's instructions, Mr. Gouges did not use the 'kicked out' joints, which were not necessary. And he stated that he had no concerns about any of the piping that was used in the well.

Mr. Gouges explained that on the day before the ejection of the pipe from the well, he had successfully set a packer tool in the well, but he noted that there were problems releasing the packer. Mr. Doit instructed Mr. Gouges to pull up on the piping with 115,000 pounds of pressure, which he did, but the packer still did not release. Mr. Gouges shut operations down at the well that evening because it was getting dark. When PowerSour informed him that it had decided to 'flow the well' and 'evacuate [the] nitrogen blanket,' Mr. Gouges expressed his concern about the hydrogen sulfide 'coming to the surface' and affecting the equipment and piping. Mr. Gouges told Mr. Doit that he 'didn't like the idea' of flowing the well, and Mr. Doit 'was well aware of his concern.' Nevertheless, Mr. Doit told Mr. Gouges 'that's what [PowerSour] wants to do, that's what we're going to do.' Mr. Gouges explained that he did not argue with Mr. Doit because he 'felt comfortable with what we were going to do because our pipe was still, you know, well within its limits to pull what we were going to pull.'

The next morning, Mr. Gouges returned to the well, and Mr. Doit told him that PowerSour intended to 'pull up' on the piping using up to 144,000 pounds of pressure, in increasing increments of 2,000 pounds of pressure. Mr. Doit also told Mr. Gouges that 'the engineers ran the numbers and everything is going to be just fine.' Mr. Gouges began pulling and, when he pulled using between 128,000 and 130,000 pounds of pressure, the piping parted at Joint 18.

Mr. Gouges explained that Snubberlovers snubbing unit had operated properly and no one from PowerSour had told him that they had a reasonable expectation that the pipe might part or break during the operation. However, PowerSour's Paul Pitting admitted during cross-examination that PowerSour had a concern that the pipe 'could possibly separate' as a result of pulling up on it. Mr. Gouges also noted that PowerSour's own completion procedures had a provision 'requiring a nitrogen pad' and the pad was to be 'applied to the wellbore before any tubing was run into the well' and was to be there before Snubberlovers 'started the hole.' And he noted that PowerSour's completion procedures did not allow for the nitrogen blanket 'to be taken off of the well while running equipment in the well.' After hearing the evidence, the jury found that PowerSour's negligence caused the accident."[8]

Operations Analysis

If pipe appears damaged or looks questionable for any reason, do not run it, even if it was just API inspected. Evidence was presented that the workstring was damaged from prior use. Photographs showed "significant indentations." The subbing supervisor kicked out a few of the joints for "deep scarring." However, evidence and testimony indicated the Oil Company snubbing expert and Company Man both inspected the pipe, but did not express concern regarding the deep gouging. Based on testimony, a Boss in Houston said to run all the pipe because it had been API inspected. The problem is that the Boss was not on location to visually assess the pipe to make that decision.

Regarding the nitrogen blanket and H_2S, it is undisputed that the workstring broke as a result of sulfide-stress cracking. Pipe can easily fail at less than 100% of book rated tensile strength because book values are determined in air, not under actual well conditions. If a well has H_2S or pressure, the failure point can be different from the book rating. For these reasons, it is understandable that the jury found that Oil Company negligence was responsible for the accident.

"Shallow men believe in luck or in circumstance.
Strong men believe in cause and effect."
Ralph Waldo Emerson, Poet[9]

23. NEVER: Assume Location Does Not Have H_2S

Hydrogen sulfide gas (H_2S) can exist on any oilfield location. Depending on the concentration, H_2S can be deadly. If a field does not naturally produce noticeable amounts of H_2S from the reservoir, it is common for folks to let their guard down when it comes to dealing with the gas.

However, small unnoticeable amounts of H_2S can build up to dangerous levels over time in various areas and pockets on location. Since it is heavier than air, H_2S can unexpectedly accumulate to deadly levels in any tank, low-lying area, windless area, confined space, pit, during drilling, flowback, and daily hydrocarbon production.

H_2S Gas Pocket Kills During P&A Operation

Let's study a fatality from an H_2S-related incident that occurred during plugging operations. Since this accident is not well-known, names have been changed or removed out of respect to all parties involved. Key excerpts (abridged and adjusted) from the fatality report are below:

The operation was to plug and abandon a nonproductive natural gas well. Pocket Gas Corporation "was in the process of plugging the base of the well with cement and then [pumping] water into the well in order to enable them to remove 4-inch casing which was being salvaged . . . this well contained H_2S which was evident by corrosion on the wellhead . . . the reason for capping this well was due to the presence of hydrogen sulfide . . . the valve was opened and a cement plug was pumped to the base of the well . . . water was being pumped into the well churning up the mud and any gas that might have been present...water circulated to the top of the well and began to go into the trench and pit . . . (it was stated it shot out of the well) . . . the pit liner began to slide into the pit.

Paul Revere ran over to pull the liner out of the pit and secure it to the slide . . . at this time a pocket of H_2S was released. . . . Paul stood up and yelled for the other employees to run. Paul dropped to the ground and Samuel Prescott ran over to pull Paul away from the pit. Samuel stepped back and then fell to the ground. William Dawes ran over to help and he fell to the ground.

George Washington was able to crawl out of the area and get to his truck in order to drive ~0.5 miles for emergency assistance because he was unable to get cell phone reception . . . the fire department arrived . . . put on their emergency rescue air packs and proceeded to evacuate the employees. . . . Paul Revere was pronounced dead at the hospital. Samuel was in critical condition. The other employees were treated and released."[10]

Operations Analysis

Lack of preparation and awareness for H_2S contributed to this accident and fatality. Looking into specific details, there were a number of contributing factors, including:[11]

1) No signs posted at the site to warn people about H_2S.
2) Employees were not provided with gas monitors.
3) No area H_2S monitoring was in place.
4) Employees were not provided with respirators (SCBA's).
5) Employees were not trained to use respirators.
6) No emergency contingency plan existed.

The combination of mistakes proved to be fatal. In regards to emergency planning, if someone goes down due to H_2S, it pays to have a specific plan in place on how to handle the situation. Running into a cloud of gas without a SCBA to rescue someone is not an effective plan, it's a panicked reaction. Get the proper equipment on first, then attempt the rescue.

"Before beginning, plan carefully."

Cicero, Politician[12]

24. NEVER: Dismiss Minor Issues with BOP, Wellhead, or Stack

Many people in this business take the BOP, Wellhead, and/or Frac Stack for granted. They think nothing can go wrong. Therefore, this equipment does not get much attention and time is not invested to become intimately familiar with these critical components of oilfield operations.

Proper installation is paramount and oversight is key. You must have a copy of the installation procedure. Missing a step or operating without a detailed installation and testing procedure can result in problems during drilling, completions, and/or production. Additionally, you must have a detailed diagram and understand all the components.

Should an issue occur with the wellhead and/or frac stack, your options are limited, especially if operating under pressure. However, if the problem is caught quickly, it can usually be addressed without a full-blown incident. You must have eyes on the BOP, wellhead, and/or Frac Stack at all times during operations. Potential problems must be identified and addressed immediately.

When inspecting the wellhead, pay attention to anything unusual. Even a drip leak must be addressed immediately. During fracing operations, the entire frac stack must be clearly visible at all times. Covering the stack in the winter months to prevent freezing has resulted in small leaks not being noticed, until they became big leaks ultimately resulting in loss of well control.

Leak Below Master Valve Turns Into Blowout

During fracturing operations, a pump hand noticed a leak at the wellhead below the lower master valve. The leak was on the C-section and was small but more than a drip. The pump hand reported it over the radio to the frac van. The Frac Treater told the Company Man and

Engineer in the frac van that the wellhead was leaking. The job was otherwise going well and currently staging to 6 ppg sand concentration. They were almost finished with the stage, with 20 minutes of pump time remaining. The Company Man exited the frac van and went to get a closer look at the wellhead. He returned to the van and told the Engineer that it did not look like a bad leak but was an issue. The Engineer mentioned that the job was almost finished and that he really wanted to get 100% of the sand pumped.

> ***TACTICAL TIP***
>
> Never frac without the ability to see the wellhead from the frac van.

Fifteen minutes later, during the flush stage, the leak got progressively worse. The Pump Hand reported it to the frac van and the information was relayed to the Company Man and Engineer. They were discussing the issue when disaster struck. The frac stack parted at the lower master valve flange.

Operations Analysis

The moral of the story is, **don't take chances with the wellhead**. In situations like this, you must shut-down operations and initiate an investigation. A drip leak on the wellhead may just be condensation, or it may be the indication of a serious internal failure.

25. PREPARE: For Weather's Impact

Weather can have a significant impact on oilfield operations. It seems every year, when the big seasonal temperature swings occur, a critical piece of equipment breaks. If there is a weakness in the system, the weather will find it. Quick temperature changes, freezing rain, windstorms, flooding and other weather events put a stress on most oilfield equipment. If there is a weak connection, there is a good chance it will be exposed.

The best way to address weather risk is to prepare for it. Check all connections, especially before the big seasonal weather changes. Winterize equipment prior to first freeze. Check and reseal electrical equipment prior to the rainy season. Tie down equipment and mobilize wind walls before windstorms. Since weather conditions are always changing and mother nature likes to keep things interesting, always keep an eye on the weather.

Ice Blockage + Pressure = Accident

Let's review an incident where cold weather was a contributing factor. Key points are summarized below:

"It was 10°F and snowing. . . . The crew was running surface casing and filling up the line with drilling fluid to weight down the casing to get it to go down easier. As reported through interviews, the safety cable (whip check) was not in place. The fill up line was connected to the 4" kelly hose with a relief valve. The 2" fill up line on the other end was in the hole about 5 feet. The rig hand opened the 2" fill up line and the pump was kicked in to fill the hole with drilling fluid.

The rig hands and casing operator did not stand clear of the pipe at the time the Driller kicked in the pump. The casing operator was leaning over the hole to listen to the pipe. Employees were unable to confirm that the 4" relief valve was open. . . . An ice blockage somewhere in the line caused the line to build pressure to 499 psi. The fill up hose blew out of the hole and was observed whipping around. . . . The end of the hammer union which was not secured hit the casing operator in the face and head...The Company Man called 911. The rig hands applied first aid. . . . The helicopter was unable to fly due to poor weather. The ambulance took 45 minutes to arrive. . . . The operation had a 24 hour stand down.

The investigation found that it was normal operation to fill the hole without a safety cable or secure it to prevent unsafe movement... the Supervisor stated they have always done it that way; they secure the long string with a swedge but not the surface casing. [After the 24 hour stand down], a new procedure was implemented to secure the hose with a swedge."[14]

Operations Analysis

Freezing temperatures caused an ice blockage to form somewhere in the line resulting in pressure building up, ultimately blowing the fill-up hose out of the casing. Although the weather was a key aspect of the root cause of this accident, not securing the hose with a swedge was also a factor.

Depending on where in the world you work, being prepared for first freeze can mean the difference between life and death. Temperatures can drop quickly and unexpectedly, causing a laundry list of problems in the oilfield. Consider constructing a seasonal weather-related preparation checklist to ensure your operation is winterized or prepared for other weather related events. If you forget or stuff slips through the cracks, a weather-related train wreck is certainly in your future.

"Men argue; nature acts."

Voltaire, Historian[15]

26. READ: Material Safety Data Sheets

Information is power to protect yourself and your company. If you are asked to handle any materials, before you do so, obtain and read the material safety data sheet (MSDS) for the product. If the material is in its original packaging, the process is fairly straight forward:

1) Look for the material label and read it.
2) Obtain the relevant MSDS for the product.
3) Confirm the MSDS product trade name has the same name as what is on the label of the product you are about to handle.
4) Read all information in the MSDS.
5) If you have any concerns, get your company chemical expert on the phone and/or call the phone number on the MSDS.
6) Utilize all required personal protective equipment (PPE).
7) Only handle the material if you can do it safely.

If you are being asked to handle or work with a material that is out of its original packaging, there is a good chance it is mixed with other materials. This makes the situation harder to assess because you must know all of the mixed products and if there are any issues, chemical changes, or increased risks when they are mixed together.

Let's consider the following scenario. You show up on location, and the Company Man asks you to clean up a liquid that spilled on the pad. In order to perform the job, you need to obtain a bunch of information which the Company Man may or may not be able to provide. Consider the following process:

1) Ask what is in the fluid and request all MSDS.
2) Read the MSDS for all chemicals in the mix. If someone tells you that a certain fluid is not dangerous, you must confirm it with the MSDS. Confirm for yourself. The person telling you

that there is no risk may not have read the MSDS, or they may not even have the MSDS.

3) Next, check the MSDS for compatibility, stability, and reactivity between chemicals, as well as handling recommendations.
4) Look at the fluid. Is it bubbling or is a gas coming off of it? If so, it may be in the middle of a chemical reaction. You probably do not want to handle the fluid.
5) Safely test the fluid to obtain more information. If you work with fluids in this business, it is a good idea to have the ability to determine pH – a measure of the acidity or basicity of an aqueous solution. You can buy pH strips (they are inexpensive) if your company does not provide them, or does not have a digital pH meter for you to use.
6) If concerned, or if you are dealing with an unfamiliar mixture, call your company chemical expert and discuss the situation with them before moving forward with the operation.
7) Utilize all required personal protective equipment (PPE).
8) Handle the material only if you determine that it can be done safely.

Drilling Mud Eats Skin Off Man

"Dutch worked for Chemical Brothers, 'a company which specializes in cleaning mud tanks at worksites for oil and gas drilling operations.' Dutch arrived at Rave Oil Corporation's worksite that day, and the EDM Company Man represented to him that the mud tank 'contained water-based mud.' The Company Man did not inform Dutch that the mud tank 'contained large quantities of caustic materials.' Based on that representation, Dutch entered the mud tank without proper safety equipment and waded in the mud.

Dutch was exposed to the caustic materials, and as a result his clothing disintegrated, large portions of his skin became severely burned, and parts of his skin fell off. Jerry, in an attempt to treat Dutch, poured vinegar on Dutch's burns, which exacerbated his injuries.

Dutch sued Rave Oil Corporation, Drum & Base Drilling, Chemical Brothers, and Jerry. . . . Dutch amended his lawsuit . . . to name EDM as a defendant. In relevant part, Dutch alleges that EDM acted negligently by, among other things:

1) Failing to inform him of the caustic materials in the mud tank.
2) Failing to provide him with adequate safety equipment for use in the mud tank.
3) Failing to properly train and supervise its employees to respond to caustic burn injuries or other medical emergencies.
4) Failing to hire, train, or supervise competent and qualified employees."[16]

Operations Analysis

If someone tells you that a chemical is not hazardous, and you are going to enter a tank full of that chemical, personally verify that it is actually non-hazardous before entering. Read the MSDS, test the chemical, and ask plenty of questions. Wear the proper PPE before performing the work, or if the proper protective equipment is not available, do not perform the job. If the Company Man gets angry and threatens to find someone else, let him do that. Protect yourself, your Company, and the quality of your life. It's not worth it to get injured.

If you are a Company Man, be careful regarding what you ask people to do. Make sure you fully understand the risk involved in a task before you ask someone to be exposed to that risk. Ensure people have the proper PPE before they begin the job, and if they do not have the proper PPE, do not let them do the work. The Company Man position holds a lot of responsibility and can expose that person to legal action if something goes wrong on location, regardless of what caused the accident.

Cementer Sits On Bucket and Dies Painful Death

After a hard day of work, the cement crew was talking with the Company Man and Oil Company Engineer on location. They were by the water tanks. During the conversation that lasted an hour, one of the Cementers sat down on a bucket to rest his legs. What he did not know was that he was sitting on a bucket of powerful chemicals.

There must have been chemical residue on the lid because on the drive home, he started to feel a burning sensation on his backside. The next day he felt very sick and was rushed to the hospital by his wife. The doctors could not figure out what the problem was. They brought in many medical specialists. Unfortunately, after five days, he died in the hospital. It is believed that the chemical entered his body via the anus when he was sitting on the chemical bucket on location.

Operations Analysis

Chemical hazards must be taken seriously in this business. Do not lean on a tote full of chemical or sit down on a drum of chemical. In fact, before you sit anywhere, look at what you are about to place your body on. Check the area to ensure it is clean and safe. Protect your body, it's the only one you have. Additionally, chemicals must be kept in a secure place, away from areas where people work or congregate. Not only could someone decide to sit on a bucket of dangerous chemical, but a person could also trip over a chemical bucket and spill it on themselves and/or on the ground. Heavily trafficked areas are not areas where chemicals should be kept.

27. CONFIRM: You're Rigged Up on Correct Well

Rigging up on the wrong well happens more often than you think. There are a number of tactics that can be employed to minimize the risk of this happening to you. Below are a few suggestions:

Well Verification Tactics

1. Install a plate with the well name and number on a piece of equipment that cannot be removed. A cellar-welded nameplate installed when the cellar is installed is a good option. You want the plate installed before drilling begins and as soon as possible after the drilling location is staked.
2. Always have a Wellsite Schematic that indicates the planned location of all wells on the pad. Once the wells are drilled, obtain an updated and confirmed schematic.
3. Run a marker joint (short joint) on each well on the pad at different depths; at least a 200' MD difference from all other wells on location. The depth of difference must be significant so you can easily confirm that you are rigged up on the correct well.
4. When running in the hole for the cement bond log and/or plug and perf operation, confirm the location of the marker joint matches the casing tally for the well you should be on.
5. Once confirmed, mark the frac stack with the well number (painted as big as possible on the stack) to minimize confusion during mutli-well operations.

"In the oil business there are few 'little' mistakes."

James Kinnear, fmr CEO Texaco[17]

4,500 feet of Lateral Lost Due to Rig Up Mistake

Thimblerig Operating drilled three wells from the pad. One well "Tall Lady #1HX" was an extended 9,500 foot lateral. The other two wells "Game of Monte #2H" and "Bent Corner #3H" were 4,500 foot lateral single section wells. The plan was to first frac the two section wells by alternating stages between wells, "zipper frac" style, then the extended lateral would be fractured alone.

The first stage on each well was pumped out the toe sleeve. Then Thimblerig Operating transitioned into plug and perf operations on the section wells. Within less than a week, the first two wells were stimulated, with 25 stages each. The crew rigged up on the third well, fractured the first stage out the toe sleeve, then ran in the hole with the plug and guns and got unexpectedly hung up at 14,500 feet MD, about 4,500 feet in the lateral section. After multiple attempts to pass 14,500' MD, the Company Man called the Engineer and explained the situation. After a lengthy discussion and investigation, the team concluded that they accidentally fractured half of the extended lateral well, Tall Lady #1HX, thinking it was one of the section wells. The reason they could not pass 14,500 feet was because they were rigged up on one of the section wells.

Once the news reached the District Manager, a meeting was called to determine how best to stimulate the 4,500 feet of wellbore that was accidentally bypassed due to rigging up on the wrong well. Reservoir engineering and geology demanded that they stimulate the bypassed portion. The operations team did not want to take the risk, but since they were responsible for the mistake, they remained relatively quiet. As with everything else in this business, the loudest voice wins; therefore, reservoir and geology got their way. The decision was to frac Tall Lady #1HX via coil tubing. During the process, coil got stuck and parted. After multiple attempts to retrieve the coil tubing, the decision was made to produce the well as is. The bypassed section was reached from the other side with a new extended lateral called Three Card Monte #1HX.

Operations Analysis

Several mistakes were responsible for this train wreck situation. When the wellhead B-section was installed on all wells on the pad and well name stenciled on the wellhead, the crew labeled the wellheads in an identical order to all wells in the area. The problem was that on this particular location, it was not correct. Additionally, the casing crew ran the marker joint within 10 to 20 feet of each other on all three wells on the pad, making it difficult to use the marker to identify which well they were rigged up on. Furthermore, the Engineer did not send the location schematic to the company men and the company men did not ask for it or double check that they were rigged up on the correct well. The entire team and service company teams assumed the wellheads were marked correctly when installed.

"If there's more than one way to do a job,
and one of those ways will result in disaster,
then someone will do it that way."
Ed Murphy, Engineer[18]

28. ALWAYS: Utilize a Safety Factor

If you have gotten this far in the OSG, using a safety factor is a no brainer. However, some people still enjoy pressing their luck, similar to playing the 80's game of chance game show called *"Press Your Luck."* In the game, unlucky timing and circumstance can result in landing on a *"Whammy"* and losing all of your money. The whammy cartoon creature would show up, take your money, then usually get blown up. For luck, contestants would chant, *"Big money, big money, no whammy, no whammy."* If you don't use a safety factor or don't use one correctly, you might as well chant *"Big money, no whammy"* because you are pressing your oilfield luck. Let's review a legal situation that helps convey the message.

Stuck Frac Port Leads to Blowout

"While Exceed Energy Partners was attempting to open a stuck frac port sleeve by applying various levels of pressure, a 7-inch piece of casing ruptured downhole in the well, causing the top casing joints and wellhead to be ejected into the air, and allowing a flow of gas and well fluids to the surface that could not be controlled.

The parties dispute whether the 7-inch casing broke apart because Exceed Energy Partners exceeded the maximum allowable casing pressure for this operation. Because of the uncontrolled flow of gas and well fluids to the surface, the well was 'out of control,' as that term is defined in the [insurance] policy, giving rise to Plaintiffs' claims for coverage. . . . To contain the flow, Exceed Energy constructed pits, and the runoff was hauled away by trucks. Cleanup and snubbing operations continued for an extended period. They later plugged and abandoned the well, and ultimately redrilled a replacement well.

"When you want to know how things really work,
study them when they're coming apart."
William Gibson, Writer[19]

Exceed Energy Partners incurred costs and expenses:

1) In attempting to regain control of the well, including plugging and abandonment ("P&A") costs.
2) Redrilling a replacement well.
3) Cleaning up pollution resulting from the blowout.
4) Oilfield equipment owned by others that was damaged.

Exceed Energy Partners gave proper notice and submitted their losses for reimbursement under the policy. Acting through internal adjuster, Oilfield Insurance Company assigned independent energy loss adjusters to investigate the claims. . . . They issued a preliminary report to Oilfield Insurance, finding Exceed Energy may have violated the policy's 'due care and diligence' clause during the fracturing job, either by using too much pressure causing the 7-inch casing to separate, or by failing to attach the wellhead to the well's 13-3/8 inch casing rather than the 9-5/8 inch casing. Oilfield Insurance then hired a Petroleum Engineer, to further review information regarding the cause and/or nature of the well control incident. The Petroleum Engineer ultimately concluded that the cause of the blowout was Exceed Energy Partners use of excessive pressure on the 7-inch casing which caused the casing to fail, and the blowout to occur.

Relying on the Petroleum Engineer's report, Oilfield Insurance denied coverage on their control of well (including P&A), redrill, pollution cleanup, and care, custody and control claims on the grounds that Exceed Energy Partners engineering decision to exceed maximum safe fracturing pressure violated the 'due care and diligence' clause in the policy. The denial of coverage letter stated in relevant part:

Despite the planned limitation of the wellbore pressure, Exceed Energy Partners elected to exert higher pressure on the 7-inch casing and it appears that this higher pressure caused the failure of the casing and the blowout ensued . . . Oilfield Insurance views the repeated pressure excursions far in excess of the designed pressure tolerances… as a failure to exercise due care and diligence in the conduct of the frac job. For that reason, Oilfield Insurance is denying Exceed Energy Partners claim.

Oilfield Insurance denied coverage for P&A and redrill costs on the additional basis that the policy did not cover these claims because the well was lost due to the pressure operation and not from any unintended and uncontrolled flow that followed the breaking of the casing. As stated in the denial of coverage letter:

There is an additional coverage issue with respect to the P&A and redrill portions of the claim. Well control coverage under Section 1A of the Policy is for expenses 'as a direct result of' the well getting out of control. Similarly, redrill coverage is afforded under Section 1B of the Policy where redrill expense was incurred 'as a result of' a well control incident. In this instance, it appears that the necessity of P&A and redrill was due to the parting of the casing, and not due to the well control event. Significantly, the casing parted before - not as a result of - the blowout. Therefore, even without the due diligence issue referenced herein, there would be no coverage for P&A or redrill expenses that resulted from the parted casing.

Oilfield Insurance engaged an expert, Mr. Stress to determine if Exceed Energy Partners exercised due care and diligence in its operations. . . . The expert's central opinion is that no reasonable operator would have exceeded a minimum safety factor of 10% below the published burst pressure for the 7-inch casing and that Exceed Energy Partners should have even further reduced the maximum pressure because of the possible wear on the casing and unknowns about the presence of fluid in the annulus. Mr. Stress also opined concerning industry safety factors, gas in the annulus, wear caused by rotational drilling, higher burst rates in high collapse casing, and considerations concerning casing couplings. Mr. Stress used a caliper log to conclude that the casing was worn and looked to industry standards to determine that a safety factor was exceeded. Mr. Stress testified that he is 'intimately familiar with the engineering standards in the oil and gas industry, and with standards applied in the area where this well was drilled.' As to the methodology used, Mr. Stress states in his initial and supplemental report:

My methodology was to review the documents and testimony, apply sound engineering principles and calculations, [and] apply industry-standard safety factors as set forth in American Petroleum Institute (API) publications, to determine the cause of the incident. I relied upon my experience in oil and gas operations, which includes engineering design, field supervision, office supervision and executive management; my knowledge of industry accepted safety practices; and my knowledge of and experience with the Texas Railroad Commission rules and regulations. All of my opinions and conclusions are made within a reasonable degree of engineering certainty, based on industry standards and my experience.

A jury trial was held. . . . The jury returned a verdict that Oilfield Insurance Company did not 'prove by a preponderance of the evidence that Exceed Energy Partners failed to exercise due care and diligence in its well operations.' "[20]

Operations Analysis

If you have safety mechanisms in place and subsequently run into a situation in which you want to do something that would require removing those safety mechanisms, think twice. From the available information on this case, it appears they set a maximum pressure for the frac job, using a pressure safety factor applied to the casing book value burst pressure. Then they had difficulty opening a stuck frac port while staying below the calculated maximum frac pressure. To see if a higher pressure would help open the stuck port, they decided to exceed the originally set max pressure and go to a pressure beyond a safety factor of 90% of the published burst pressure for the 7-inch casing.

The operation resulted in the top casing joints and wellhead being ejected into the air, causing a blowout. The insurance company denied coverage on the grounds that Exceed Energy Partners' decision violated the 'due care and diligence' clause in the insurance policy. The case went to trial and, although a jury determined that the insurance company did not prove that Exceed Energy Partners failed to exercise due care and diligence, there are several aspects regarding this accident that should be discussed.

A safety factor should always be utilized, but it must be utilized correctly. Not using a safety factor correctly is just as dangerous as not using any safety factor because both can result in disaster. Evidence in this case was introduced indicating the casing had wear. Mr. Stress used a caliper log to conclude that the casing was worn. Although there is uncertainty based on available information, it is possible this casing was drilled through, prior to the frac, which caused the wear.

Fracturing down casing that has been drilled through can pose additional risks when compared to fracturing down brand new casing that has never been drilled through. If you are going to frac down casing that has been drilled through, depending on the situation, consider running a log to determine the extent of drilling-induced wear before the frac job. This information can be used to determine the safe max treating pressure for the frac job.

There was also uncertainty on the presence of fluid in the annulus. Depending on the situation and calculations, it is generally a good practice to ensure the annulus is full of fluid and is monitored throughout the operation. If applying backside pressure is necessary, remember that pressure on the annulus does not cancel out pressure on the casing. There are additional stresses involved. Safety factors should be tailored for each individual environment, dependent on well conditions. Applying a blanket 80% safety factor on every well can get you into trouble.

"There has been at least one lawyer engaged for every barrel of oil that ever come out of the ground. You might wonder if they pay so much to lawyers how do they ever make anything out of the oil? Foolish question."

Will Rogers, Actor[21]

29. RUN: The Numbers

Complex oilfield operations generally require certain calculations to be performed prior to initiating activities. It is safer and more cost effective to determine an operation has a high probability of failing with proven equations, engineering principles, and computer programs compared to executing an operation, causing an incident and then having someone show you mathematically, after the fact, that you had no chance of success. It is very embarrassing and potentially devastating if an injury or fatality is involved. Additionally, not performing the proper calculations could be considered negligence, or gross negligence, by a government regulator or in a court of law.

Sometimes people forget or dismiss the importance of performing calculations. Other times, the field crew assumes the calculations were already performed by an engineer in the office. Unless you have a copy of the calculations, never assume they were performed. Assume they were not performed and ask for the calculations. If working with a service provider, ask for a copy of the computer program to run various scenarios. Change the variables and assumptions to see the effect on operations. Understanding this additional information before operations are started gives you options in case things do not go exactly as planned. For example, knowing how much weight you can pull, max pressure you can apply, max differential stress that can be withstood and how the combination of all forces and stresses impact equipment, safety, and risk – before beginning operations – puts you in a position of power to deal with adversity.

> ***TACTICAL TIP***
>
> When you are given calculations for an operation, always review the assumptions and input data that goes into the computer program or equations. Often this information needs to be updated to reflect actual field conditions.

"Do many calculations lead to victory, and few calculations to defeat."

Sun Tzu, The Art of War[22]

Lack of Calculations Contribute to Multiple Deaths

Not performing the proper calculations contributed to the death of seven people during a snubbing operation. The U.S. Chemical Safety and Hazard Investigation Board (CSB) incident report, regarding the snubbing catastrophe, is reproduced below (key excerpts):

"The day started at 7 a.m. with a safety meeting. . . . By noon, they successfully snubbed-in about 250 ft of tubing string composed of:

A. 20 ft blast joint
B. Four joints of tubing
C. Two 10 ft pup joints
D. 20 ft blast joint
E. Joint of tubing
F. 10 and a 4 ft pup joint
G. 20 ft blast joint in the BOP stack

After lunch, they made up the next section consisting of:

H. 6 ft long pup joint
I. Packer
J. On-off tool
K. 10 ft pup joint

> ***TACTICAL TIP***
>
> Never snub anything that cannot be supported by engineering calculations, especially jewelry.

It was noticed after assembly that the position of the 6 ft and 10 ft pups was reversed from the positions shown on the tubing arrangement chart prepared by the Exploration Company Design Engineer. It was decided by the Packer Representative, the Drilling Supervisor, and the Snubbing Supervisor that the position of these pup joints was not important, and that the section would be snubbed as assembled.

This section was made up on the ground under the direction of the Packer Representative, hoisted into place, and then the joints were made-up tight using tongs. The traveling slips were secured on the 6 ft pup joint just below the packer, and the string was snubbed down several feet until the traveling slips were on top of the stationary slips. The traveling slips were then secured about 21 inches above the on-off tool.

The Snubbing Operator became concerned when he saw that the workstring was bowing excessively when he started to apply pressure to it. He commented that such a short (only 6 ft long) pup joint was under the packer and that there were potential problems with this snubbing arrangement. He suggested reversing the positions of the 6 ft and 10 ft pup joints. His concerns were brought to the attention of his Supervisor and the Exploration Company Drilling Supervisor. As a result, his Supervisor went up into the workbasket to discuss the operators concerns.

Precisely what occurred in the next few minutes is unclear. Several witnesses were interviewed, but their recollections varied slightly. It is known that the snubbing proceeded with no changes in the positions of the components in the tubing string and that the Forman was in the workbasket when the snubbing operation was resumed.

When the stationary slips were released, or shortly thereafter, all of the tubing string below the traveling slips (about 350 ft) blew out of the well with such explosive force that a loud bang was heard. The shock wave from this energy release was sufficient to knock several of the people in the area off their feet and propel them away from the wellhead. The natural gas flowed freely out of the well forming a dense vapor cloud that ignited moments later by a spark probably generated when some of the tubing hit the rig. A vapor explosion and fireball resulted which in turn ignited new gas as it was roaring out of the well. The individuals that had been knocked down earlier, and were starting to get up, were knocked down again. The flames from the well reached about 20 ft above the rig. The five individuals in the workbasket and two others on the rig had almost no time to react. There was only a time interval of two to three seconds between the time that the tubing started to come out of the well and the vapor explosion.

Those that were able, ran from the area and assembled at the predetermined emergency action meeting place. Emergency responders were promptly notified by means of a cellular phone. Seven people died in the explosion; four more were injured, all from burns. Falling tubing broke a piping manifold about 300 ft from the wellhead releasing additional gas. This leak was stopped before it ignited, by closing an isolation valve."[23]

Operations Analysis: OSHA Investigation Findings

"The OSHA investigation began in response to a fire and explosion that occurred. . . .OSHA alleges that [the companies] were negligent in providing specific and adequate methods of prompt emergency escape from the work basket of a rig-assisted space-saver type hydraulic snubbing unit. Additionally, [they] **did not perform the necessary calculations** to determine the imposed forces and applied stresses to the pipe section that was being pushed or snubbed into the hole. The forces exerted on the pipe caused it to buckle resulting in it being ejected out of the well by the natural gas pressure causing a blowout and subsequent explosion and fire."[24]

Operations Analysis: CSB Investigation Findings

"Three tools were intended to be inserted, almost one right after the other. A blast joint (a heavy walled section of hardened pipe) had already been partially snubbed-in. A packer (used to seal the annular area between the well casing and the tubing string) and an on-off tool (quick disconnect coupling) were setup and ready to be snubbed-in. The clamping force of the slips could damage either the packer or the on-off tool. It was therefore necessary to place the slips in such a position that they engaged only the short section of tubing between the blast joint and the packer, and the tubing above the on-off tool.

When the traveling slips above the on-off tool were engaged, and the jack started to push the tubing into the well, the tubing buckled and broke. This removed the force that was restraining the tubing string in the well and allowed the tubing to be blown out of the well. This permitted gas to rapidly escape from the well and form a vapor cloud, which ignited. The men working in the basket at the top of the rig and those on the lower platform had no possibility of escape from this series of events, which occurred in only a few seconds."[25]

Key Findings

1) "The person operating the snubbing equipment, his supervisor, the factory representative for the packer and on-off tool, and four others that were working on the rig were killed in the incident. Four additional support people were injured. Only three of the workers on site at the time were not injured.

2) The workstring (the portion of the tubing string that was snubbed-in) failed initially in the threaded portion of the mandrel (inner pipe) on the top of the packer. This section was initially subjected to excessive compressive forces followed by sufficient tensile and shear forces for it to break. This type of failure is consistent with buckling failure.

3) The workstring buckled and broke because the length of tubing below the traveling slips had an unsupported length that exceeded the critical buckling length for the force needed to push the workstring into the well.

4) No operating procedures could be found in the Corporate Snubbing Procedures Manual that detailed the snubbing procedure to be used when items of different diameter and physical properties were being used.

5) No training procedures could be found in the Corporate Training Manual that detailed the snubbing procedure when

items of different diameter and physical properties were being used.

6) **No calculations could be found that determined the critical buckling length, or conversely, the maximum snubbing force for a given length of tubing, for the particular arrangement of tools in this operation.** The snubbing company however, has assured the investigation team that calculations for the original tubing arrangement and well pressure had been completed and approved using their normal approval process. The company has stated that the only copy of these calculations was in the possession of the Snubbing Supervisor.

7) No evidence could be found that safety accessories and procedures commonly used by the oilfield industry were used to help prevent an incident from occurring.

8) No evidence could be found that there was a procedure for ensuring good communications between the four or more parties involved in the design, supply and snubbing of the tubing string during the design and specification of tools for the tubing string. Nor could evidence be found that the contractors work was checked to verify that it met the needs of the client for this specific application.

9) No evidence could be found that the oilfield tool industry has, or makes available to their customers, the data needed to determine the resistance of a workstring containing their tools, to buckling. . .

10) Recommended practices for this type of work are available from some organizations, but the level of detail provided in this information is not adequate to prevent an incident like this from occurring."[26]

CSB Investigation Root Causes

1. "Operating procedures were inadequate:
    a. The operating procedures make no mention of how to control buckling when various tools or multiple diameter tubing is included in the tubing string to be inserted into the well. No calculations could be found that determined the critical buckling length for the tubing and tools being inserted into the well.
    b. Project management procedures are inadequate.

TACTIC VIOLATION

Core Tactic #29
RUN: The Numbers

TACTIC VIOLATION

Core Tactic #7
ALWAYS: Have a Written Procedure

2. Training procedures were inadequate:
    a. The operating manual, while not as detailed as it should be, did describe several steps and procedures that are to be followed to minimize the potential for buckling. However, these procedures were not followed. It is unclear what training the individuals had prior to performing this work. But it is clear, that they either had not learned during the training, they had forgotten what they had learned, or that they had insufficient training in the first place.
    b. A shorter section of tubing was used between the blast joint and the packer than had been specified in the tubing string design. This exacerbated the buckling potential by positioning a blast joint in the snubbing BOP, thus increasing the cross sectional area at that point which increased the upward force that had to be

overcome by the snubbing unit, thus increasing the potential for buckling. No evidence could be found that this change had been reviewed to ensure that it could be done safely.

c. The actual wellhead pressure was much higher than that used to design the tubing string. However, no evidence could be found that the effect of this change on the snubability of the tubing string was evaluated.

d. The training procedures do not cover how to perform portions of the work that is required by the operating procedures to be performed.

3. Communication between the parties involved was inadequate:

   a. No evidence could be found that the owner of the well communicated to the snubbing contractor or the tool provider, the operating conditions within the well. This information has a significant impact on the selection of the tooling to be provided and how it is to be snubbed-into the well.

> ***TACTIC VIOLATION***
>
> Core Tactic #5
> **KEEP: Constant Effective Communication**

4. The evidence indicates that the well owner delegated the responsibility for snubbing to the snubbing contractor but did not review their work plan to ensure that it would be workable and fully meet their needs. In addition, they delegated the responsibility for the specification of the tools to the

> ***TACTIC VIOLATION***
>
> Core Tactic #1
> **ALWAYS: Protect Yourself and Your Company**

provider of the tools, but did not review their work to ensure that it could be safely snubbed-into the well.

5. Both the well owner and the snubbing contractor noticed that the **wellhead pressure was higher than expected** when they arrived at the site. However, no evidence could be found that either party reviewed this change to determine its impact on the job, or received approval to proceed with the job under these conditions. It was not recognized that **this change made it impossible to snub-in the entire workstring.**

> ***TACTIC VIOLATION***
>
> Core Tactic #9
> **MAINTAIN: Situational Awareness**

> ***TACTIC VIOLATION***
>
> Core Tactic #11
> **ALWAYS: Pay Attention to Detail**

6. Management in both the exploration company and the snubbing company, suggested that brine be added to the well to reduce the well pressure. The exploration company supervisor in the field however, chose to ignore this advice and to proceed with the job."[27]

"It does not do to leave a live dragon out of your calculations, if you live near him."
J.R.R. Tolkien, Writer[28]

30. PROVIDE: Solutions

If you don't solve problems at work, you don't work in the oilfield. Everyone who works in this business deals with problems every day, especially those who work in operations, also known as the "real world." Younger folks learn early on that it is not wise to bring problems to the boss without potential solutions. Don't be afraid to ask for help. However, when an issue occurs, consider potential solutions so that when discussing with colleagues and superiors, you have something to offer during the conversation and problem-solving process. Always be open to ideas and potential solutions from others regardless of title, department, or experience level.

Never hide problems. Some folks try to, particularly when there are underlying communication or personal issues. This is not a good practice, since it usually comes out or is found out at some point. Once people realize that information was withheld, the trust level quickly deteriorates. Additionally, if an issue is withheld, people will assume that the person hiding the issue caused the problem and did not want to get in trouble.

Some people are excellent at finding problems, which is a good thing. However, once they find the problem and report it, for some reason they think the job is done. Finding and reporting problems is encouraged, so you should not hesitate to report an issue even if you don't have a solution. At the same time, we are employed to find and solve problems. Providing solutions is part of the job. Talking through problems with colleagues and your own personal network of contacts and industry experts is a great asset to solving problems. Consider a short story on the next few pages that helps illustrate this point.

"The problem-solver has significant value in all times; a person who can only identify the problem – not so much. And frankly, nobody likes a bearer of only bad news. So pair a problem with a solution; that way, you'll be remembered and appreciated."

Martin Craighead, CEO Baker Hughes[29]

Guns And Plug Hung Up In Frac Stack

During a routine multi-stage horizontal shale completion, after fracturing the 1st stage through the toe sleeve without issue, the wireline crew rigged up to run the plug and guns for stage 2 – the first plug and perf stage on the "Toretto" multi-well pad location in Texas. As they attempted to run in the well, the plug and guns got hung up while trying to pass through the frac stack and well head. Monty Carlos, Company Man, calls Engineer Edward F. Fairlane.

Company Man: "*We have a problem on the Toretto #1H well. Wireline getting hung up in the stack somewhere.*"

Engineer: "*This is the first P&P stage, right?*"

Company Man: "*Yes, gonna try and wiggle it through. You okay with that?*"

Engineer: "*Yes, but be careful and not rough with it. Make sure the Wireline Engineer is okay with this also.*"

Company Man: "*He also suggested it. I did it successfully on another well. It might work. We are doing it now. . .*"

Engineer: "*Wellhead and frac stack issues are serious business, you know what I mean. We should not be having this issue. I wonder if the stack is level or the lubricator is at an angle.*"

Company Man: "*We just got through the stack and wellhead, running in the hole. Might have caught a lip between the stack and wellhead.*"

Engineer: "*On my way, be on location in 30 minutes.*"

Mr. Fairlane walks over to Dwayne Diesel's office, Director of Operations, before driving to location.

Engineer: "*On my way to location to help solve an issue. We are having some problem getting through the stack with the plug and guns on the Toretto #1H. Just worked it through so we are running in the hole right now, but I wanted to be on location to see the situation first hand."*

Director: "*Good deal, let me know what you find.*"

Later that day on location, inspecting guns on surface after stage 2.

Engineer: "*No marks on the guns. Stage 2 frac'd good, no issue. Maybe not an issue.*"

Company Man: "*Bad news, just got hung up on the other well, Toretto #2H.*"

Engineer and Company Man walk over to the #2H. Wireline Engineer joins the discussion at the wellhead.

Wireline: "*You want to work it through again?*"

Engineer: "*Raise the assembly into the lubricator and lower it back down again slowly.*"

Wireline: "*Okay, boss.*"

As they slowly lower the plug and guns into the frac stack, they hear what sounds like a metal-on-metal clap. Wellhead Tech, Mr. Tanner, walks over.

Company Man: "*Metal-on-metal, catching on something.*"

Engineer: "*Measure the distance.*"

Wellhead Tech: "*Might be casing collapse.*"

Engineer: "*On both wells right at the surface? Maybe, but I don't think so. Tanner, anything different on this setup?*"

Wellhead Tech: "*No, same as always. Here is a schematic.*"

Company Man: "*Let's wiggle it through.*"

They proceed to work it through, but this time they really have to yank on the wireline before they get it through the stack and wellhead.

Engineer: "*That did not look good. It's not safe. We are not going to do that again. No way.*"

Company Man: *"You think the casing is collapsed?"*

Engineer: *"I hope not and I don't think so because this problem is happening on both wells. Let's measure the diameter of everything on the plug and guns."*

Company Man: *"We did that already. The plug has the largest diameter."*

Engineer: *"Then what, these centralizers?"*

Company Man: *"Yeah, then the setting tool."*

Engineer: *"When is the last time we ran these centralizers? I think this is the first time, right?"*

Company Man: *"Yeah, first time. Do we really need them?"*

Engineer: *"I don't think so. Do we need this?"*

Wireline: *"No, I never run this crap. But we are going to have to disassemble everything."*

Engineer: *"We are not going to jam the plug and guns again, it's too rough. A little wiggle is okay, but not getting all medieval on it, you know what I am saying, right, you agree?"*

Company Man: *"It's not good. The plug could get damaged then pre-set or the guns could detonate, who knows."*

After removing the centralizers, the assembly passed through both wells without issue. Edward calls the boss, Mr. Diesel.

Engineer: *"We removed the centralizers, everything is good now. We never run these things anyway. Our setup is probably not compatible. The centralizers are very rigid, need to bring them into the shop and test on passing through a mocked-up frac stack and wellhead, unless you absolutely want to run them?"*

Director: *"We don't need to run those things right now, extra jewelry. Let's finish this frac and move to the next location."*

"Simplicity is the ultimate sophistication."

Leonardo da Vinci, Engineer[30]

VII

TROUBLESHOOTING

Solving problems is part of the job in this business. Some people, usually those who never worked in the field, think that if everything is planned and perfectly executed, there should be no problems. This logic is flawed because the business of Earth exploration is working on rocks miles beneath the surface, millions of years old, not homogeneous, and not constructed by man. There are many aspects about the Earth that we do not understand. Although if anyone on this planet knows anything about the secrets of the Earth, by God, it is those of us in the oil and gas business.

Every day you go to work, be prepared to have problems and solve problems. Things you never thought were going to happen are going to happen. Having the ability to identify a problem and address it quickly to move forward with the operation is a valuable skill. At a high-level, it can help to have a systemized tactical approach to troubleshooting oilfield problems.

The foundation for successfully troubleshooting oilfield problems is to be prepared for anything. Experience helps: if you have successfully dealt with a similar problem in the past, you can employ a field-tested solution to the issue at hand. However, obtaining that experience can be expensive, since dealing with problems in this business is, in and of itself, very expensive.

"No problem can withstand the assault of sustained thinking."

Voltaire, Historian[1]

Troubleshooting Oilfield Problems

1	ALERT: To Potential Problems
	Vigilance is rewarded in the oilfield. Catch an issue before it turns into a train wreck. Reaction speed is critical. Showing up on location unprepared or horsing around, not paying close attention to detail is a recipe for disaster. Sooner or later, you will be eliminated by the Earth or by the Man.
2	VERIFY: Problem Exists
	Since oilfield problems can be deceptively difficult to identify, once signs indicate there may be an issue, a verification process should be initiated. Collect facts and data.
3	IDENFITY: Cause
	Identifying the cause is part of the troubleshooting process that can be stressful if you are not prepared. Tactical methods to troubleshooting are provided on the next page.
4	CHOOSE: Potential Solution
	Choose wisely. The potential "solution" employed can increase or decreased your options going forward. It pays to think 2 steps ahead.
5	DEPLOY: Potential Solution
	Knowing what to do and doing it correctly are two different things. For example, reading a book on brain surgery does not make you a brain surgeon. Know when to ask for help.
6	CONFIRM: Problem is Fixed
	Never assume what you implemented actually fixed the problem. Always confirm the issue has been addressed before moving forward.
7	SHARE: Experience
	Documenting and sharing the learnings from your troubleshooting adventure helps to prevent a potential train wreck from occurring to someone else. After all, sharing is caring. However, if you only care about yourself, say nothing, do nothing.

"While the masses are waiting to pick the right numbers and praying for prosperity, the great ones are solving problems . . .
the bigger the problem you solve,
the more money you make."

Steve Siebold, Speaker[2]

Consider the following tactics when troubleshooting problems:

Tactical Troubleshooting Methods

I	Scenario Plan
	Develop a selection of "what if" tactical options/solutions, taking into account various combinations of uncertainties, which you can deploy when encountering problems. (covered in Chapter 9)
II	Catalog Common Problems
	Construct a reference book of frequent problems that have been encountered in your location and area of operation. Reviewing historic well reports and talking with as many people as possible can help you create your own, job-specific, survival guide.
III	Collect Facts
	When troubleshooting oilfield problems, fact, fiction, and opinion can get all mixed up. Before long, a theory is presented by someone involved in the operation. If the person is confident and convincing, when everyone else is uncertain and confused, theory can erroneously become fact. Then an action is taken, often resulting in a train wreck.
	To avoid this scenario from playing out, carefully collect as much data and facts as possible. It helps to write everything down. List all the facts, numbered in order of importance. Disseminate facts, data, charts, records, and photographs to all those involved, especially management and engineers. As new information is learned during the troubleshooting process, update all those involved. Establishing an email thread and/or group text can help reduce the burden of notifying everyone individually.
IV	Discuss with On-location Expertise
	Know who the experts are on-location. Take the time to interact with your team to learn their strengths and weaknesses. When troubleshooting, always let the team on-location know what is going on. Communication is critical. You never know who has dealt with a similar situation or knows someone that can help determine a potential solution.
V	Notify Your Boss and Engineers
	Always notify your boss and engineers when troubleshooting a problem. In many situations, they may not be on-location. However, with communication technology available today, there is no excuse not to keep leadership in the loop. There is value in getting feedback and help from people that are not clouded by the stress of being in the middle of the action on-location.

VI	**Engage Service Providers**
	In many cases, the problem you are troubleshooting will involve service providers. Engage them to help and ask them to engage their network of experts. Often, within service companies, there is an expert who has seen a similar problem.
VII	**Consult Expert Contacts and Mentors**
	Construct a contact list of people whom you know are experts in specific areas of oilfield operations. This list should include folks from across the industry and across the country. This is your "Phone-A-Friend" option and should be used without hesitation. Get multiple opinions; do not reply on one person's opinion no matter how confident they sound.
VIII	**Challenge Theories and Proposed Solutions**
	Proposed theories and solutions should never be accepted without being vigorously challenged with the facts. If someone presents a theory as to why the current problem exists, even if they have 50 years of experience, do not assume they are correct. If you are concerned, speak up.
IX	**Think Two Steps Ahead**
	Every action taken can increase or decrease your future options. Certain actions, originally thought to be solutions, can end up trapping everyone on location in a train-wreck-type situation. Never underestimate the downside risk. Before employing a potential solution, determine the worst case scenario if you are wrong. What will you do? What are your options?
X	**Run the Numbers and Test on Small Scale**
	Perform proper calculations to see if the theory or proposed solution works mathematically. If it is possible to test a theory or proposed solution on a small scale, consider doing it before implementing on the well itself, which can be very expensive or life limiting if you are wrong.

"When you start looking at a problem and it seems really simple, you don't really understand the complexity of the problem. Then you get into the problem, and you see that it's really complicated, and you come up with all these convoluted solutions. That's sort of the middle, and that's where most people stop . . .
But the really great person will keep on going and find the key, the underlying principle of the problem — and come up with an elegant, really beautiful solution that works."

Steve Jobs, Apple[3]

VIII

CORE TACTICS: 31 - 40

Oil and gas companies, from a strategic and structural perspective, have trended toward specialization on both the services side and operator side of the business for many good reasons. Since almost everyone is segmented and specialized within the industry, the value of communication is much more critical today than ever before.

Without excellent communication between departments, undesirable situations occur. Part of the communication problem is that people are segmented and specialized to such an extent that they do not have a working knowledge of what people do outside of their area of expertise. It's hard to have strong communication if you do not know what needs to be communicated. For this reason *Core Tactics: 31-40* continue to be a mix of knowledge from across the oil and gas business.

"If we assist the highest forms of education – in whatever field – we secure the widest influence in enlarging the boundaries of human knowledge."

John D. Rockefeller Sr., Billionaire Oilman[1]

31. KNOW: The Rock

Please do not be confused. I am not suggesting that you get to know the wrestler and actor, Dwayne "The Rock" Johnson. Studying his movies will most likely not help you in the oilfield. Unfortunately, most folks in this business know more about The Rock's movies than they know about the reservoir rock or overburden formations. Rock ignorance is incredibly unforgiving, disrespectful to your chosen profession, and vengefully offensive to the Earth – an entity which always plays for keeps.

If you ask the guys on location to tell you about the rock, and they start talking about Mr. Johnson's latest movie, you might have a problem with your operation. However, the field team is not entirely to blame. Often this critical information is hoarded by the technical staff and not shared with field personnel implementing the work. Some office folks do not realize that the people on location must have this information to make informed decisions, troubleshoot problems on location, and prevent incidents from occurring.

For example, in certain formations a 0.05 ppg change in mud weight can mean the difference between gaining and losing drilling mud, potentially contributing to an undesirable situation. This type of detailed rock-related information must be disseminated to the guys in the field. The best place for this information is in written procedures and/or emails that are sent to everyone on location on all shifts. Do not assume people inherently know this information or are sharing information on location. Often, interpersonal relationship issues between service providers, company men, and superintendents exist, resulting in a breakdown of communication.

"If I see the things I want to see – good high resistivity, good porosity, and good permeability, like the three cherries spinning into view at the end of a slot machine play – then we've got a well, a producer."

Rick Bass, Geologist[2]

If you work in the field, do not expect rock related information to be sent to you or be included in the procedure. If you do not have this information, do not hesitate to ask for it. You need it to make informed decisions. Especially during troubleshooting. Below is a list of critical rock-related information that can help avoid troubled operations.

Critical Rock Properties

- Temperature gradient and bottomhole temperature (BHT)
- Pore pressure for all zones encountered
- Hydrogen Sulfide (H_2S) content
- Corrosive intervals
- Water-bearing intervals
- Formation tops
- Lithology and characteristic changes
- Frac gradient for all zones encountered
- Depth of potential faults
- Depth of potential lost circulation intervals
- Formation dip and structures
- Production potential (oil, gas, water rates)

Vice President Hoards Rock Info Causing Train Wreck

Seismic data indicated a potential fault at the end of planned horizontal well, Vacuum Boom #6H. The technical team's plan was that if early indications during drilling confirm the validity of the seismic interpretation, they would TD the well before crossing the potential fault. All of the technical information and potential plan was left out of the procedure because the division Vice President felt that it was too sensitive. He did not want the field guys to have access to the data due to concerns it would fall into competitor hands.

It was a busy time of year. The technical team was simultaneously preparing for a quarterly internal executive management meeting. The geologists and engineers overseeing the drilling of the well were not paying close attention, as the drilling operation was unknowingly confirming the validity of the seismic interpretation. As they approached the fault, seepage losses turned into partial losses which progressed to severe loses. By the time the technical team realized what had happened, they had already drilled across the fault and lost complete returns. The well then went on a screaming vacuum. Cement was pumped on multiple occasions in an attempt to fill the fault and circulate cement into the intermediate casing.

At this point, they were over budget by 30%. Finally, circulation was regained and cement was pumped again. It was thought that enough cement was circulated into the curve. The job was initially considered a success. Everyone was in such a rush to get off location that the field team pulled the BOP off the wellhead without checking for pressure on the backside. Gas quickly covered location and caught fire. A well control team was called in and was able to obtain control within a day. No one was injured or killed in the accident.

Operations Analysis

If the field team was notified of the potential fault, this train wreck would have been prevented. The Geologist wanted to include the interpretation in the prognosis sent to the field, but was overruled by the VP. Initially, the Geologist and Drilling Engineer were blamed for the disaster, as it was determined they were not paying close attention to the job. The management meeting was a distraction.

The Vice President planned to fire both of them. The VP spoke to Human Resources to get the paperwork together, and Human Resources decided to conduct their own investigation into the disaster, which cost the company over $10,000,000. After the incident investigation, it was the Vice President who ended up getting fired over his decision to withhold technical information from the field team.

"Nature does not reveal her mysteries once and for all."

Seneca, Ethicist[3]

32. STOP: When Something Doesn't Seem Right

When something does not seem right, trust your instincts and stop what you are doing. Take a few minutes to think things through. Speak with others on location and in the office about what you are seeing. Consider calling a trusted colleague removed from the operation, with no vested interest, for a second opinion. Under no circumstances should you ignore a potential issue and hope it goes away or does not cause an accident.

Flowback Hazards Ignored Leading To Flash Fire

The well screened out on stage 38 of a 40 stage plug and perf frac job and was being flowed back to obtain the plug ball. The plan was to get the ball back, flush the well of sand, and then pump the plug and guns down for the next stage. However, the team was having difficulty recovering the ball. After several wellbore volumes of fluid were recovered, flowback gas was beginning to overwhelm operations.

The Company Man was concerned due to a significant amount of gas venting off the flowback tanks and starting to cover location. Several pieces of equipment were running, and he had a sinking feeling in the pit of his stomach that something bad was going to happen. The crew thought the ball would be recovered any minute. However, minutes turned into several hours as the gas increased. Speaking with the flowback crew, they thought everything was okay and were not concerned about the gas catching fire because it was a wet gas and they said they had seen worse.

Flowback Boss: *"You worry too much. We do this all the time."*

Company Man: *"Don't you think we should stop?"*

Frac Treater: *"I think we are going to get the ball back any minute now. Just you see. We'll be fracing again."*

Company Man: *"You, guys, don't think this is too much gas?"*

Flowback Hand: *"No, this is nothing. You should have seen this other well 5 miles from here."*

Company Man: *"What happened on that well?"*

Frac Treater: *"We got the ball after flowing a ton of gas."*

As the Company Man and Treater went to check the flowback tanks, a flash fire occurred across location, burning a dozen people. The well was shut-in and the fire quickly burned itself out. Several people were taken to the hospital with injuries. The Company Man, Treater, and Flowback Hand all received 1st degree burns on their face and hands.

Operations Analysis

When you get that feeling something bad is going to happen, it's time to take action. Sensing danger is a valuable skill. Do not dismiss or ignore it. Develop your oilfield intuition rather than suppress it as the Company Man did in this situation, resulting in multiple people getting burned. The Company Man should have shut the job down.

Additionally, the Flowback Hand and Treater should have taken the Company Man's concern seriously, as opposed to talking him out of his concern. Just because they were able to flowback more gas on an offset well does not make it a safe procedure. They got lucky on the other well, which led them to believe the current situation was not dangerous. If equipment is not able to handle an operation (a large amount of flowback gas in this case), stop the operation and address it. It is not worth putting the team in harm's way. Injuries take a long time to heal, much longer than the inconvenience of stopping an operation.

"Get all the education you can, but then, by God, do something. Don't just stand there."

Lee Iacocca, Ford Mustang[4]

33. EXAMINE: Equipment Prior to Use

Brand new equipment can fail. In many cases, new equipment carries a higher risk profile because it's new. All the bugs have not been worked out, and it has not been field-tested. Additionally, do not assume equipment, new or old, is free from design defects. Manufacturing oilfield equipment is not immune from engineering flaws or fabrication mistakes. Just like anything else that is built by humans, it is not a perfect process. Defects on an assembly line in a far off country can find their way to an oilfield near you, resulting in an incident.

Manufacturing Defect Causes Paralysis

"The crew had just pulled the 52nd stand of rods. While the Derrick Man was positioning the rod stand into the rod fingers on the rig, the rod elevator came off the rod hook and fell below (approximately 60 feet) to the work floor area striking the Floor Hand in the back. The injured Floor Hand was airlifted to the hospital sustaining severe central spinal cord stenosis T11-T12, right 8 through 12 rib fractures, T3 through T11 spinous process fractures, and scapula fracture. Due to the severity of injuries to his vertebral column and spinal cord, the result was paraplegia from the waist down. The investigation concluded the rod hook latch had a manufacturing defect."[6]

FACT

4,700,000 American workers are injured on the job each year.

National Safety Council[5]

Underrated Ball Valve Results In Rig Fatality

Let's review another equipment-related incident. This oilfield fatality involves utilizing an underrated low pressure ball valve. Key excerpts (abridged and adjusted) from the fatality report are as follows:

"Iron Services was using a 600 psi ball valve in a 5,000 psi triplex drilling mud pump to pump water. The Driller was assisting in getting a water line in position to run water down the casing when the 600 psi ball valve failed at a line pressure of 3,072 psi. The end of this line struck the Driller in the head causing fatal injuries. A 2nd employee was injured in the accident."[7]

Operations Analysis

In both incidents, equipment issues played a critical role. If the equipment were carefully inspected for potential problems prior to initiating operations and corrective action taken, both incidents would most likely have been prevented.

Regarding the first incident, *"Manufacturing Defect Causes Paralysis,"* the rod hook latch had a manufacturing defect. The defect caused the rod elevator to come off the rod hook, fall 60 feet, and hit a man below, resulting in his paralysis. In this case, careful inspection of the latch, including function testing, could have been key in identifying a defect prior to initiating operations. Action must then be taken by removing the defective equipment and notifying your company and the manufacturer that there is an issue. This way, everyone can be made aware of the problem.

The second incident, *"Underrated Ball Valve Results in Fatality,"* is the result of a low pressure valve installed on a high pressure pump without effective pressure controls. Examining the pressure rating of all components, relative to maximum system pressure via a pop-off or other pressure limiting device, is a task that should occur prior to pressuring up any system. Checking the pressure rating on each piece of equipment may be laborious but life on location depends on it. Never assume everything is rated to the same working pressure. Often, components are installed without taking max pressure into account. It is also not unheard of for someone to shut a valve while pumping.

34. ALWAYS: Maintain Good Housekeeping

Keeping equipment clean is important not only in terms of appearance and brand value but also in terms of safety. If equipment is kept clean, it is easier to identify signs of potential failure, including:

- Leaks
- Loose connections
- Misalignment
- Stress fractures
- Corrosion
- Debris buildup
- Incorrect rig up/construction
- Missing components
- Lack of lubrication
- Equipment damage
- Excessive wear

If you can see that there is a potential issue, you can address it. Additionally, good housekeeping is critical in the cab of all vehicles on location, as the following story illustrates.

Soda Bottle Causes Blowout

Mister Energy Slickline Services was performing a routine pressure/temperature dip-in for Fountain Oil Operating, which had identified a correlation between bottomhole temperature and well performance. This led to discovering a geothermal hot spot which defined the overpressured core area in a new shale play. To map the asset, the geology department requested accurate bottomhole temperature data on all wells across the field. Fountain Oil engineers employed Mister Energy Slickline Services to run memory gauges on 100 wells. Mister Energy was the low bidder on the project known for underbidding and then overcharging. They also were known to not keep their equipment properly maintained or care about good housekeeping. Their trucks were a mess and since their company acronym abbreviation also happened to spell out the word, everyone called them "The MESS."

The untidy service company was rigging down and moving to the next location when one of their vehicles driven by a man they called Pigpen lost control, striking the wellhead. Water and gas immediately covered location. Everyone evacuated, including Pigpen, who was able to get out of the truck, which was stuck on top of the wellhead. A fountain of flames soon followed and a well control company was called in. Fortunately, no one was hurt in the accident or operation to control the well.

Accident Analysis

During the accident investigation, Pigpen claimed the boom truck brakes stopped working. He stated that he pressed the brakes over and over again to stop the truck but they would not work. Since millions of dollars were lost and Fountain Oil was in a legal situation with Mister Energy Slickline Services and the insurance companies, a thorough mechanical forensic analysis was performed on the burned truck recovered from the blowout.

They found the driver side of the cab filled with empty root beer bottles and energy drinks. Pigpen loved drinking root beer, so this made sense. He also never cleaned the truck, hence the nick name "Pigpen." When the investigation team disassembled the brake system, they made an astonishing discovery. One of the empty soda bottles was wedged under the brake pedal arm assembly. The team concluded it got wedged in there at some point before the accident and prevented the brake pedal from being able to be pressed down.

"Things are not always what they seem; the first appearance deceives many; the intelligence of a few perceives what has been carefully hidden."

Phaedrus, Poet[8]

35. MONITOR: Corrosion and Erosion

Nothing lasts forever. Even the Sun has a life span. Scientists predict the Sun will last another 5 billion years, using up its supply of hydrogen and helium, then collapse swallowing the Earth and a few other planets in the process.[9] Oilfield equipment will also not last forever. Although it's not the Sun that swallows you up, but the power of the Earth in the form of corrosion and / or erosion.

Preventative measures must be taken to minimize the impact of corrosion and erosion. Monitoring equipment is key. Consider employing a carefully planned preventative maintenance program. You must know when certain components need to be fixed or replaced before they fail. Do not wait until the equipment fails to replace it. That would be a dangerous and unforgiving tactic.

Unnoticed Corrosion Contributes To Accident

"The operator was performing routine scaling operations in an attempt to repair a leaking surface controlled subsurface safety valve (SCSSV). The operator decided to perform an acid job to reduce the amount of scale around the SCSSV after several attempts to remove the scale with a wireline unit were not successful. The operator pumped approximately 100 gallons of 1% HCL into the well and allowed it to soak overnight.

The day after the acid job they re-entered the well to perform more scraping. A seal ring on the bottom flange below the master valve began to leak and dry gas was released into the atmosphere. Since the SCSSV was not operable and the leak was below the master valve, the operator was not able to prevent the escape of natural gas. The platform was evacuated without injury."[10]

"Negligence is the rust of the soul that corrodes through all her best resolves."

Owen Felltham, Writer[11]

Operations Analysis

The investigation determined there were several causes of the accident, including a corroded ring gasket. "The exact condition in which the gasket was in prior to the incident is not known; however, it is known that a ring gasket downstream was replaced two days prior due to a possible corrosion failure. Pictures taken during a post-mortem inspection of the tree indicate the gasket involved in the incident was heavily corroded prior to the incident. Therefore, the amount of corrosion on the ring gasket which lead to the loss of mechanical integrity was concluded to be a cause of the incident."[12]

The key warning sign occurred two days prior to the blowout when a corroded ring gasket was replaced downstream of the ring gasket that failed. When that ring gasket was identified and replaced due to corrosion failure, a red flag should have gone up, bringing into question the integrity of all other ring gaskets in the system, the most critical being the seal ring below the lower master valve. This was especially critical, considering an acid job was going to be performed.

Determining the condition of equipment, especially the wear and tear from corrosion and erosion, prior to initiating operations is important when managing risk. A strong preventative maintenance program with corrosion monitoring integrated into the process would have reduced the risk of this incident occurring.

"Time will rust the sharpest sword,
Time will consume the strongest cord;
That which molders hemp and steel,
Mortal arm and nerve must feel."

Sir Walter Scott, Poet[13]

36. CHECK: For Flammable Gases and Liquids

Natural gas is odorless. It is a misconception that you will always be able to smell natural gas or any hydrocarbon. People are often confused between natural gas and H_2S which has a distinct rotten egg smell at low concentrations. However, H_2S is not present in all natural gas. Therefore, in many cases, there is no smell. Additionally, natural gas for household use contains an additive called mercaptan, giving natural gas a smell, so you can smell it if there is a leak. Mercaptan is added to natural gas to make it smell. It's considered an odorizer. However, mercaptan is not added to the gas on oilfield locations. This makes detecting natural gas on oilfield locations very difficult without a multi-gas detector.

Oil and natural gas produce a large amount of energy, and since we all work in the oil and gas business, sooner or later you will find yourself surrounded by significant quantities of natural gas and oil. Under the right conditions, it only takes a small spark to witness the full power of these hydrocarbons. Even static electricity generated on location can ignite natural gas under the right conditions. In a number of deadly oilfield accidents, hot work provided the ignition source, as the following story and fatality report illustrates.

Double Fatality From Welding Oil Tank

"Employees were directed to weld on a 210 bbl tank containing approximately 85 bbl of a flammable oil and gas mixture. . . . Upon initiation of the welding operations, the tank exploded resulting in two fatal injuries."[14] The site was in a remote location and consisted of "the wellhead, piping, separator, dyer, and 4 tanks."[15] The OSHA investigator was approached by the fire marshal and local law enforcement to speak with an individual that had information on the accident. The individual stated to OSHA that "He was called by the Engineer…that the leaks that he repaired, were leaking again, and that he was going to have [another person] weld the tank but they were having a hard time locating the 3 seeping holes. He stated that he asked the Engineer, 'What tank?'"[16] When told the name of the well, "he stated to the Engineer, 'Do not put fire to that tank.'"[17] He stated to the Engineer, "He would not weld on a tank that contained oil. He also added [that on] the day of the explosion he spoke with the Engineer and advised him 'Not to put fire to the tank.'"[18]

Operations Analysis

Apparently, the 85 barrels of oil could not be removed by "conventional means" because of the position of the drain valve. A company "removed the crude [oil] to about 1 foot from the bottom due to [the] position of the drain valve. The welder said that the oil was below where he would be welding."[19]

This operation is, of course, totally unacceptable. There is no excuse to weld on a tank that contains volatile materials, regardless of how difficult it is to remove those volatile materials. Sometimes, it is not easy to do the right thing, or it may cost more. In this case, it was apparently not easy to remove all of the oil. That is not a valid excuse to continue with operations. There is no valid excuse. Do it right, or don't do it until it can be done correctly.

The investigation concluded, "The employees were not trained and instructed in the safe procedures to be undertaken prior to the initiation of the welding operation to include, but not limited to inspection of the tank to determine if it was emptied, cleaned, and did not contain any volatile materials. . . . Prior to the initiation of the welding operations, the tank was never inspected by supervision to determine its contents and if the contents presented a hazardous condition."[20]

"Education is not the filling of a pail, but the lighting of a fire."
William Butler Yeats, Poet[21]

37. WATCH: For Excessive Vibration

Oilfield equipment is designed to withstand specific conditions that can vary significantly, depending on the piece of equipment and operation it is engineered for. Excessive shaking, vibration, twisting and collisions can cause equipment to fail abruptly, especially if the equipment is also under internal pressure. If a collision occurs with a piece of equipment, or if it is dropped, the equipment could be damaged and fail at some point in the future unexpectedly, even though visually the equipment looks brand new.

Additionally, if a piece of equipment has a working pressure of 10,000 psi, depending on the equipment, the pressure rating may not include vibration or other forces that could act on the equipment. If you apprehensively notice a piece of equipment shaking, that is not designed for vibration, it is safe to assume it is excessive.

Shaking Breaking Blowout

Let's review a court case in which vibration allegedly contributed to causing a well control situation on a frac job. Key excerpts (abridged and adjusted) from court documents are below:

"This action stems from an alleged loss of control at Super Shale's wellsite during hydraulic fracturing operations. Force Wellheads specifically alleges that during the fracturing operations, Frac Rocker Services experienced problems and was apparently unable to maintain a constant, steady flow of fracturing fluids through the wellhead into the wellbore and reservoir at the desired pressures and rates. Frac Rocker Services knew that it was experiencing problems during its fracturing operations, but continued with the fracturing process rather than shutting the operation down as a reasonable and prudent completions contractor would, and should, have done. . . .

Frac Rocker operations moved, shook, vibrated, rocked, swayed, tilted and/or torqued the wellhead, causing the flange at the bottom to experience loads and stresses in excess of the tolerances specified by the American Petroleum Institute (API) for this connection. Frac Rocker operations separated the flange, allowed fracturing fluids to escape from the wellbore and caused Super Shale's alleged damages. . . . Frac Rocker fracturing operations misused and abused the equipment in the wellhead, causing the flange at the bottom to experience forces in excess of API specifications and tolerances. Frac Rocker Services is responsible to Super Shale for the defects alleged to exist in the wellhead because the flange would not have separated but for Frac Rocker acts and omissions during the fracturing operations...

The Court found that while the wellhead was integrated into the hydraulic fracturing operations system, the wellhead and the hydraulic fracturing system were two separate products provided by two separate independent parties. For Frac Rocker Services to be responsible to Force Wellheads for indemnification, Frac Rocker Services would have to have manufactured the wellhead and supplied it to Force Wellheads, who in turn supplied it to Super Shale. In this instance, that was not the case and, therefore, since Frac Rocker Services neither manufactured nor supplied Force Wellheads with the wellhead, no legal relationship exists between the two parties."[22]

Operations Analysis

Based on the available legal documents, it appears that during the frac job, Frac Rocker Services frac pumps had trouble maintaining constant rate. The pumps may have lost prime, were knocking, shaking, or had some other issue which placed an undesirable force on the equipment, causing the wellhead to vibrate and twist. A wellhead flange failed as a result, causing a blowout.

Although hindsight is 20/20, as soon as the issue started to impact the wellhead – one of the most critical components in terms of well control – the frac job should have been shut down. It is difficult to quantitatively measure excessive vibration. However, it can be determined by visually monitoring the wellhead. After performing a few frac jobs, you will know what the wellhead should look like during pumping. If the wellhead starts to shake violently, you have to shut down. Based on the situation, you may have to replace wellhead components. Shaking, vibration, and torque will damage the wellhead. Replacing frac stack and/or wellhead components in the middle of a frac job is a tough decision to make due to the cost and inconvenience. I assure you, no one will be happy to hear that the wellhead needs to be replaced, but sometimes it's the prudent thing to do.

Early in my career, I was working as a field engineer on the services side of the business. We were pumping a frac when the Line Man radioed into the frac van to tell us that it looked like the frac stack and wellhead were bending. I looked out the van window, and although it was not clearly evident, if you looked at the stack carefully, it did appear that the wellhead was tilting over. We asked the Company Man if he wanted to shut down. He told us to keep pumping, which we did. By the end of the frac (single frac on a vertical well), the stack and wellhead were bent over 45 degrees. It was luck that the stack / wellhead did not separate from the casing and cause a blowout.

Looking back at both situations, the frac job should have been shut down as soon as it was evident that the wellhead experienced undesirable loads and stresses. Once pumping ceased, careful inspection of the stack and wellhead would have provided a good opportunity to identify that the stack and wellhead were damaged. Most likely, this conclusion would have led to the decision that the operation could not continue until the equipment was fixed or replaced.

"If you want to find the secrets of the universe,
think in terms of energy, frequency and vibration."
Nikola Tesla, Engineer[23]

38. NEVER: Frac Without a Pressure Relief System

During fracturing operations, many different processes are occurring simultaneously. Troubleshooting issues during the job is not uncommon. Pumps go down, you lose proppant concentration, your fluids may not be ideal. The one thing you absolutely cannot allow to occur is to exceed maximum treating pressure.

Exceeding the designed maximum treating pressure is a surefire way to destroy well integrity, damage surface equipment, or cause a serious accident. If you are responsible for a fracturing operation, exceeding maximum allowed treating pressure is also a good way to lose your job.

Some seasoned hands think they can catch a screenout or shutdown the pumps in time before blowing anything up, if a quick screenout occurs. That may be true, but it only takes a few seconds for a pressure spike to occur. Additionally, if a hydraulic valve is accidently closed, you will not be able to react fast enough. Furthermore, in the age of multi-well fracturing operations, people have been known to close the wrong well in by closing the active manual value during a fracturing operation. Hard to believe, but it's true.

There are frac crews that depend on electronic kick-outs. If you are on one of these crews, you need to understand that kick-outs will not prevent the pressure from exceeding the pressure at which you set the kick-out at. Electronic kick-outs only kick the pump into neutral or shut the pump down; they do not immediately release the pressure. Therefore, if you set your kick-outs at 10,000 psi and someone shuts a valve downstream of the pumps when you are pumping, the pressure can easily exceed 10,000 psi.

If you are a customer and you tell the service company to set max treating pressure at 10,000 psi, do not assume that they are using a full open full bore immediate pressure relief system to prevent the pressure from exceeding 10,000 psi. Unless you specifically ask for it, you should assume they are using electronic kick-outs and/or a spring loaded pop-off. A spring-based pop-off may not fully release pressure. It will open but the spring system may not be full open full bore and may not be able to deliver immediate pressure relief. Therefore, pressure can exceed the pop-off set point, sometimes significantly exceeding maximum pressure. Always choose a pressure relief system that fits your specific application and situation.

Max Frac Pressure Exceeded Parting Casing

During routine fracturing operations at low pressure (3,000 psi treating), pressure quickly increased above 10,000 psi. Electronic kick-outs were set at 9,000 psi but this pressure was exceeded as there was no immediate pressure release system rigged up. The casing parted at the top joint, and 13,000 psi was on the annulus which also did not have a full open unloading valve to provide immediate pressure relief. The annulus had a spring-type pop-off. The annulus wellhead B-section was only rated to 5,000 psi and it quickly failed. A large uncontrolled release occurred, as there was communication between the formation and annulus since it failed at the wellhead. Fortunately, the reservoir was normally pressured, so a hydrocarbon release did not occur and the well died after a few hours. The reason for the pressure spike which caused the incident was that the gel system lost cross-link on 5 ppg proppant concentration.

Operations Analysis

It is critical to understand your pressure relief system not only on the main line but also on the backside. Take the time to study how things may unfold if max pressure is exceeded. Understand how everything is rigged up and where the weak points are in the system. Run stress and load programs to determine how all aspects of the operation impact contributing forces on the casing or tubing. Often, determining maximum pressure is not as simple as looking up the burst pressure on the casing that you are pumping the frac down. There are many other factors involved. Below is a list of key variables that you must know to determine maximum treating pressure.

Max Pressure Determination Variables

1) Surface equipment ratings: *frac lines, frac stack, wellhead.*
2) Tubular quality (e.g. Wear from drilling through the casing or corrosion from old age or a corrosive subsurface interval).
3) Tubular burst and collapse pressure: *casing and/or tubing.*
4) Tubular diameters and weights: *casing and/or tubing.*
5) Body yield strength and joint yield strength.
6) Young's Modulus and poisson's ratio.
7) Depth to perforations: TVD and MD.
8) Depth to top of good cement and cement bond quality.
9) Weight set on the slips.
10) Bottomhole temperature and geothermal gradient.
11) Surface fluid temperate.
12) Slurry density.
13) Fluid parameters: *thermal conductivity, specific heat, viscosity.*
14) Pressure on backside and density of fluid on backside.
15) Frac job variables: *rates and volumes.*

TACTICAL TIP

Casing weight set on the slips is often overlooked during the pre-frac analysis. If excessive weight is set on the slips, you may not be able to frac the well without making adjustments.

If any variables change during the job, you must redo the calculations. For example, if frac fluid temperature drops, you may need to reduce max pressure. When running the calculations, consider flexing the variables for various scenarios to see what you can and cannot do.

"The hydraulic fracturing process is a critical technology, not only to this country, but also to the world."

George Mitchell, Shale Pioneer[24]

39. ALWAYS: Lock Out Tag Out

One hundred and sixty-seven people were killed while working on North Sea offshore production platform Piper Alpha. The root cause initiating the disaster is primarily attributable to poor company processes (Lock Out Tag Out, Permit To Work, Handover Meetings) and the negligence of multiple men, due to lack of knowledge, low work quality, poor procedure, lack of communication, no formal training, and an overall lack of attention to detail. Many additional factors, including poor design, maintenance process, safety measures, risk management systems, and emergency procedures, contributed to the significant loss of life and ultimate destruction of the platform.

At the time of the incident, 226 people were on board Piper Alpha. One hundred and sixty-five of them were killed and 61 survived, many with significant injuries. Additionally, two people were killed during rescue operations while working on a Fast Rescue Craft (FRC). The incident, full of incredible bravery and feats of man, unfolded quickly as Piper Alpha was destroyed in 50 minutes after the first explosion.[25]

Piper Alpha Disaster[26]

- Oil was discovered in the Piper field in 1973
- Reservoir covered 12 square miles
- Production started in 1976
- Piper Alpha platform was located 110 miles NE of Aberdeen
- The platform provided the ability to drill wells, produce, separate, and process reservoir fluids
- Designed throughput was 250,000 bopd
- There were 36 wells; production trees arranged in 3 rows of 12
- Piper was connected to other platforms and to land by 4 pipelines (1 oil pipeline and 3 gas pipeline)

- Platforms linked to Piper Alpha by pipeline include:
    - Claymore platform, 22 miles from Piper
    - Tartan platform, 12 miles from Piper
    - MCP-01 platform, 34 miles from Piper
- In the summer of 1988, Piper was engaged in a work program involving a number of major items, including:
    - Pressure safety valve (PSV) recertification of 300 valves
    - Preventative maintenance on Condensate Pump A
- In the first week of July 1988, gas smells, incidents, and false alarms occurred on several occasions:
    - Gas release from the gas conservation module (GCM) which resulted in a temporary evacuation
    - Various gas smells in the dive complex area resulting in the shutdown of the diving compressors
    - A leak on a nipple on Valve 17, the Gas To Claymore (GTC) platform valve. It was necessary to shut down, isolate, and depressurize the line to fix the valve
    - Pressurized line break at a production tree
    - Leak on LP switch on condensate Pump B. Switch rated below required pressure range of 0 - 700 psi
    - A number of false gas alarms on a centrifugal compressor. The plan was to change out gas detectors
- Average production for July 5th, 1988
    - 119,000 bopd (stb)
    - 7,500 bcpd
    - 33 mmcfd (flow across Piper from Tartan platform)
    - 50 mmcfd (lift gas circulation on Piper)
- Score Ltd. was the contractor hired by Occidental to perform PSV recertification on 300 valves
- Condensate Injection Pump A (see diagram) was shut down for maintenance work on the morning of July 6th, 1988

- There were 226 people on board Piper Alpha
  - 164 people on day shift
  - 62 people on night shift

Condensate Injection Pumps Diagram

PSV – Pressure Safety Valve

GOV – Gas Operated Valve

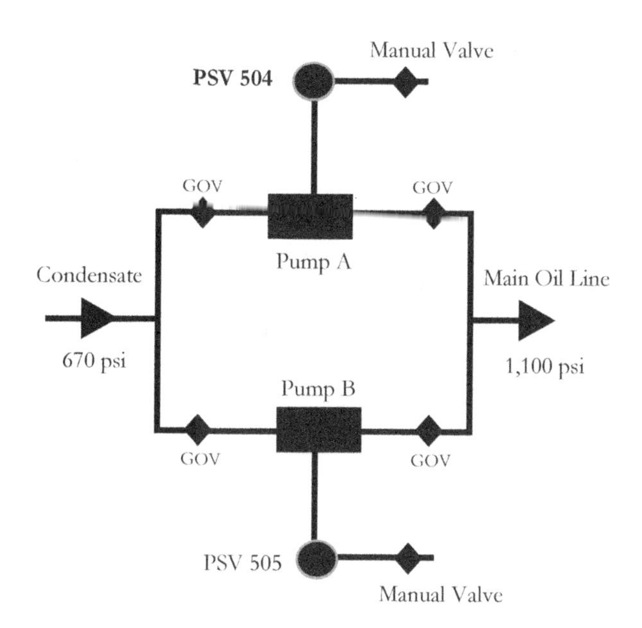

"A book is made from a tree. It is an assemblage of flat, flexible parts imprinted with dark pigmented squiggles. One glance at it and you hear the voice of another person, perhaps someone dead for thousands of years. Across the millennia, the author is speaking, clearly and silently, inside your head, directly to you. Writing is perhaps the greatest of human inventions, binding together people, citizens of distant epochs, who never knew one another. Books break the shackles of time. Proof that humans can work magic."

Carl Sagan, Astronomer[27]

Piper Alpha Disaster Timeline[28]

Factual information in the timeline is reproduced from the Public Inquiry. There is conflicting information on the events of July 6th, 1988. This is my best effort at outlining a timeline leading up to the disaster:

7:00 a.m. – Operations Superintendent asks Dayshift Lead Production Operator to take Pump A offline for preventative maintenance and put Pump B online.

Pump A is electrically isolated, suction & discharge valves closed, condensate blown down.

> ***TEST YOUR SKILLS***
>
> **What's wrong with the isolation of Pump A?**
>
> *No physical locks (with keys) are installed to Lock Out the isolation valves.*

7:10 a.m. – Score Supervisor (first tour offshore as a supervisor) and Score Mechanic, get the Permit To Work (PTW) for PSV 504 refurbishment and take it to maintenance to be signed.

7:40 a.m. – Score Supervisor takes the PTW for PSV 504 to the production office to be signed by the Operations Superintendent.

7:45 a.m. – Lead Maintenance Hand brings the PTW for maintenance on Pump A to the safety office.

8:00 a.m. – Score Supervisor goes to the Control Room to inform the lead operator and get the PTW for PSV 504 signed. He did not know who the person was or if he was the lead operator, but the man signed the PTW. He left the PTW in the Control Room.

> ***TEST YOUR SKILLS***
>
> **What's the problem with the Supervisor's visit to the Control Room?**
>
> *Lack of thorough effective communication.*

9:00 a.m. – Mechanic asks for help removing PSV 504.

10:00 a.m. – Mechanic prepares blind flanges and tools. Supervisor prepares test equipment.

1:00 p.m. – Supervisor and Mechanic have lunch.

2:00 p.m. – PSV 504 is taken down & moved to the end of Module C.

2:30 p.m. – PSV 504 is lifted by crane to the Score container.

3:00 p.m. – Score Supervisor begins work on PSV in the Score container.

3:10 p.m. – Mechanic takes blind flanges down to Module C to cover the open pipes resulting from the removal of PSV 504.

4:00 p.m. – Mechanic returns to the Score container and confirms that he fitted the blind flanges. They continue to work on PSV 504 in the container.

It is believed that the blind flange installation was not leak-tight. Allegedly, the installation was only finger-tight. A significant amount of research, analysis, and testing was performed to reach this conclusion. Additionally, there is no pressure test confirming installation integrity. The work is not confirmed or quality-checked.

5:10 p.m. – The Lead Production Operators hold a Production Operator handover meeting. Key items and issues are discussed. No mention is made regarding the status of PSV 504.

5:30 p.m. – The Lead Maintenance Hands hold a maintenance handover meeting. Day Shift Lead Maintenance Hand explains Pump A was shut down and electrically isolated but no work had started on it. The shift meeting is informally conversational in nature. Day Shift Lead Maintenance Hand does not mention anything regarding PSV 504.

5:40 p.m. – Occidental QA rep certifies PSV 504 test.

6:00 p.m. – Score Supervisor goes to the Control Room to arrange a crane to lift the PSV as it was ready to be reinstalled. The crane was not available; therefore, it was mutually agreed that the PTW should be suspended. Score Supervisor suspended the permit by writing "SUSP" in the gas test column, as there was no place on the permit for suspension. He had never suspended a permit before. Score Supervisor placed the permit on the desk.

There is uncertainty as to what exactly happened. It is likely the PTW for PSV 504 was placed on the desk because the process supervisor was in his handover meeting, and it was common practice to place PTWs on the desk and not speak to the supervisor.

TEST YOUR SKILLS

What's wrong with the handover meetings?

They are based on memory and only utilize verbal communication. No review of current PTWs, log books, or operations notes occur.

TEST YOUR SKILLS

What's wrong with the permit suspension process?

1) *There is no permit suspension process.*
2) *Worksite is not inspected to ensure conditions and equipment are safe for work to be suspended.*
3) *LOTO of suspended worksite equipment is not confirmed.*

Later that evening…

9:45 p.m. – Working "Condensate Pump B" tripped (went down). They radio the Phase I Operator to inform him of the situation.

9:46 p.m. – Lead Production Operator left the control room to check out the condensate pumps.

9:50 p.m. – Lead Production Operator returned to the control room saying, "Pump B will not restart." Pump A was down for maintenance but he wanted to get the PTW for Pump A signed off so that Pump A could be reinstated.

9:52 p.m. – The PTW for Pump A and red tags were signed off; however, there are conflicting stories on exactly how and by whom.

9:55 p.m. – Lead Production Operator went back down to the Condensate pumps to open the Pneumatic Gas Operated Valves (GOVs) to try and start the pumps.

9:56 p.m. – Lead Production Operator is at the GOVs, lining them up to start up a condensate pump. Phase I Operator is at the main control panel to help with the restart. However, he is called away and does not participate in the restart. Another two men show up to help the restart. The attempt to restart Pump B fails.

Then, allegedly, Lead Production Operator (who died in the accident) pressurizes the discharge of Condensate Pump A by "jagging," – repeated brief opening of the GOV.

9:57 p.m. – High-pitched hissing noise heard by many across platform. First gas alarms start, followed by additional gas alarms.

10:00 p.m. – Initial explosion in Module C, fueled with condensate, followed by huge fireball and fire; firewalls are destroyed.

10:01 p.m. – Pipe is ruptured in Module B; oil pool fire starts causing a huge plume of black smoke. A second fireball occurs.

10:04 p.m. – "*Mayday…explosion and fire on the oil rig on the platform and we're abandoning the rig.*"

10:06 p.m. – "*Mayday (repeated)…we require any assistance available any assistance available we've had an explosion and a very bad explosion and fire and the Radio Room is badly damaged.*"

10:08 p.m. – "*Mayday (repeated)…we're abandoning the Radio Room we're abandoning the Radio Room we can't talk any more we're on fire.*"

TEST YOUR SKILLS

What are the flaws in the PTW system?

1. *There are multiple permits and they are not cross-referenced.*

2. *There is no permit suspension process.*

3. *The permits are not centralized & organized.*

4. *Permits are dumped on the desk in the Control Room with no communication.*

Due to these Four Fatal Flaws in the permit system, they do not know about the permit and work on PSV 504.

10:20 p.m. – Major explosion due to the rupture of the Tartan gas riser causing a high pressure gas fire. A number of men jump into the sea as a result of the explosion.

10:33 p.m. – Congregation of men in the galley call out to Tharos (semi-submersible fire intervention vessel, hospital, and gangway for access to platforms):

"*People majority in galley area, Tharos come. Gangway. Hoses. Getting bad.*"

10:50 p.m. – Massive explosion due to the rupture of the MCP-01 gas riser adding to the intensity of the high pressure gas fire. The explosion destroyed a FRC, killing most of those on board. Men jump from the helideck and other parts of the platform into the sea.

10:51 p.m. – Structural collapse of Piper starts. Tharos pulls back to escape its own destruction.

11:20 p.m. – Claymore gas riser ruptures. The drilling derrick collapses. The platform tilts to the east. Men still on the platform are forced out of shelter on to the pipe deck. A number of survivors jump into the sea.

- 37% of people on night-shift died (23 of 62)
- 87% of people off duty died (142 of 164)

"I really don't know why it is that all of us are so committed to the sea, except I think it's because in addition to the fact that the sea changes, and the light changes, and ships change, it's because we all came from the sea. And it is an interesting biological fact that all of us have in our veins the exact same percentage of salt in our blood that exists in the ocean, and, therefore, we have salt in our blood, in our sweat, in our tears.

We are tied to the ocean. And when we go back to the sea - whether it is to sail or to watch it - we are going back from whence we came."

John F. Kennedy, U.S. President[29]

Methods of Escape[30]

Sixty-one people survived the destruction of Piper Alpha. There was no organized escape. A number of survivors said that it was their familiarity with the platform that saved them. Detailed knowledge of the platform was critical because the oil fire produced a significant amount of thick black smoke which covered a large portion of Piper Alpha, including the life boats making them impossible to reach. The oil fire and thick smoke combined with the high pressure gas fire made escape from Piper Alpha difficult.

- 27 descended by rope from the 68 ft level to 20 ft level
- 15 jumped from the pipe deck (133 ft free fall into the sea)
- 7 descended by rope or hose from the 68 ft level into the sea
- 5 jumped from the helideck (175 ft free fall into the sea)
- 5 jumped from the 68 ft level
- 1 walked down the stairs to the 20 ft level
- 1 jumped from a roof next to the derrick

A number of the survivors who jumped from significant height thought they would die from hitting the water; however, they held their nose with one hand, and held down their life jackets with the other arm to minimize the risk of breaking their neck when they hit the water. One man performed an amazing Olympic-type dive from 68 feet.[31]

"Then out spoke brave Horatius,
The Captain of the gate:
To every man upon this Earth
Death cometh soon or late.
And how can man die better
Than facing fearful odds
For the ashes of his fathers
And the temples of his gods."
Horatius at the Bridge[32]

Key Points

The bottom-line is that on July 6th, 1988, the Score Supervisor and Mechanic were in the process of performing routine maintenance on a condensate pump pressure safety valve (PSV 504). The Mechanic removed the PSV for maintenance and improperly installed a temporary blind flange (round flat metal plate), with only finger-tight screws, and not leak-tight. The flange was not inspected and not tested. Additionally, no lock out tag out (LOTO) equipment or any other equipment was installed to physically prevent the introduction of pressurized hydrocarbons into the temporary blind flange. Not utilizing a physical lock to LOTO equipment under maintenance was common practice on Piper Alpha.

CRITICAL FAULTS

1. Flawed PTW System
2. No Cross-Referenced Permits
3. Desk Dumped Permits
4. Lack of Communication
5. No Physical Locks for LOTO
6. Weak Handover Meetings
7. 1st Time Untrained Supervisor
8. No Work Quality Inspections
9. Finger Tight Installation
10. Deficient Training

Due to ineffective process and faults in the Permit To Work system and the shift handover, accurate information was not transmitted to key personnel regarding work status occurring across the platform. There were two maintenance jobs in progress, preventative maintenance on Condensate Pump A, and recertification of Pressure Safety Valve (PSV 504). There were two PTW issued which were not cross-referenced or connected to each other.

It was common practice to not cross-reference permits, a flaw in the PTW system on Piper Alpha. Additionally, it was common practice to dump PTW's on the maintenance lead / process supervisor's desk and not talk to the supervisor. It was also common practice for the supervisor to sign the PTW after the work had been completed before looking at the job to confirm it was done correctly and was safe to operate.

The Control Room team only knew about the preventative maintenance scheduled for **Pump A**. They knew that major work had not yet started on **Pump A**, it was only shut down and electrically isolated. Operating without accurate information, once Pump B went down, Lead Production Operator jagged Condensate **Pump A** which introduced pressurized hydrocarbons into the system containing the blind flange which was not leak tight. Hydrocarbons escaped from the flange and ignited, causing the initial explosion which kicked off a chain reaction of events, resulting in a number of additional fires and explosions that led to the death of 167 men.

"Nobody else [on Piper Alpha] had received regular and formal training in the operation of the Permit To Work system.
Everything was learning on the job.
The problem with learning on the job
is you perpetuate and
accumulate errors."
Brian Appleton, Technical Assessor[33]

40. STAY: Alert and Prepared For Anything

Getting proper sleep is critical in order to stay alert and be prepared for anything. I find that this is more of an issue with the night staff than with day work personnel. Finding a good night crew is hard. Often, folks that work the night rotation stay up during the day to do random tasks and personal chores, then show up to work without any rest. Coming to work unrested is dangerous and can contribute to an accident. When you are tired, instead of paying attention to potential issues, a lot of energy is focused on trying to stay awake.

Lack of Sleep Results in Coil Tubing Train Wreck

On the second stage of a horizontal multi-stage completion, treating pressures were close to maximum, erratic and unpredictable. The Treater, Frac Engineer and Company Man, kept on pumping proppant as though it was not an issue. They were halfway finished with the job when a screenout occurred. Although multiple attempts were made to flow the well back and displace the wellbore with clean fluid, they could not pump more than a few barrels.

Coil tubing was rigged up and proceeded to wash out to bottom. Problems with the pump and injector delayed the operation one day, resulting in additional stand-by charges on all frac and support equipment. Once the wellbore was clean and injection was confirmed at pressures below 9,000 psi, wireline was run and attempted to be pumped down. However, the well pressured up to maximum levels, and the plug and guns could not get on depth.

Wireline was pulled out of the hole and coil tubing rigged back up to TCP (Tubing-Conveyed Perforating) the next stage. The coil tubing crew did not have relief. Therefore, the same crew would be running back in the well. As a result, the coil crew was operating with only a few hours of sleep. As coil was running in the hole, the operator fell asleep in the cab. At the same time, someone accidentally shut the annulus in. Pressure started increasing and was not noticed. The Company Man had stepped out of the control cab to address other business. The differential pressure actuated the pressure operated firing head and the guns accidentally fired in the curve of the horizontal well.

Operations Analysis

The cause of many accidents in the oilfield is lack of attention to detail. Digging deeper, one may find that lack of attention to detail is due to lack of ability to focus on the task at hand as a result of lack of sleep. When you are not rested, it makes working on high-stakes operations very dangerous. Issues that you would typically catch go unnoticed until they build up into a train wreck.

In order to stay alert and prepared for anything, you must get the proper rest. It is not natural to sleep during the day and work at night. However, proper conditioning can allow your body to adapt. Some folks are better at the night shift than others. If you cannot adapt to working nights or having difficulty staying alert, request a shift change. If that is not possible, change jobs because it is not worth hurting yourself or someone else.

"No job is so important and no task so urgent that the necessary steps cannot be taken to perform it safely and maintain the health of our employees, contractors and the public."

Dave Hager, CEO Devon[34]

IX

TACTICAL SENARIO PLANNING

When dealing with adversity in the oilfield, it helps to have multiple weapons, if not an arsenal of well-planned and team-vetted problem solving options at your disposal. When troubleshooting, it is not uncommon to be required to deviate from the initial plan in order to be successful. If your written procedures are not flexible and only include a sequential process, with no optionality or consideration for potential problems, you may be exposed to an undesirable situation, should the operation unexpectedly need to deviate from the initial game plan.

SURVIVAL SENARIOS

Scenario planning provides the added benefit that once hypothetical scenarios are contemplated, preventative measures are often taken to minimize the risk of an undesirable scenario becoming a reality.

Identifying an issue and knowing when to deviate from the game plan is as hard or harder than executing unplanned actions, especially under the stress and confusion of real world field operations. Plan for adversity, at least in your mind and on paper. There is minimal additional cost in taking these actions. Being prepared reduces the chance that you will make a mistake when things are happening quickly in real time.

"Everyone has a plan 'till they get punched in the mouth."
Mike Tyson, Boxer[1]

It is easy to have a mental meltdown when multiple activities are occurring and you are dealing with conflicting information. A stressful situation combined with fatigue, poor weather, or challenging subsurface conditions often makes the best choice of action unclear.

Tactical scenario planning is a powerful risk-reducing tool you can utilize to address many of these concerns. Tactical scenario planning in the oilfield is primarily focused on developing a selection of detailed "what if" scenarios, combined with tactical options/solutions, to address the issues outlined in each scenario, while taking into account various combinations of uncertainties. Should you face adversity, these planned potential solutions are your weapons: cocked, locked, and ready to rock. Below are several implementation methods to incorporate tactical scenario planning into your operation.

Implementation Methods

1	**Flowcharts**
	A popular tool to implement scenario planning on oilfield operations is flowcharts since they utilize easy to understand, often binary, workflow diagrams. Flowcharts highlight a problem, then show a solution model with various options, visually displayed in different boxes. Three examples are presented in this chapter.
2	**Simulation Games**
	Pilots, astronauts, and the military, just to name a few, use simulation games for training purposes. Although simulation games are often implemented via scale models or video simulators, they do not have to be. A low cost implementation method is to construct oilfield storylines on paper, where each person on the team steps into a character in the story. A moderator presents the team with problems during a simulated oilfield operation. Depending on team decisions and actions, the moderator develops the story in different directions.
3	**Field Exercises**
	Implementing scenario planning with full scale field exercises is taking simulation games from the conference room into the yard or to location. Running through the motions of troubleshooting problems in the field brings you one step closer to the real thing.

4	**Decision Trees**
	Although decision trees and flowcharts look similar, they are very different tools. A decision tree is a schematic consisting of three types of nodes (decision, chance, end) connected by lines in a similar fashion as a flowchart. However, decision trees often incorporate probability, cost, and profit to different parts of the tree. This helps determine the best course of action to generate the maximum expected value. That could be decisions which result in the lowest expected costs, highest expected payoff, or some combination of both.
5	**Monte Carlo Simulation**
	Named after the casino-filled city in Monaco, Monte Carlo simulations incorporate chance and randomness to provide a range of outcomes including the probability they will occur. Almost any excel model can be easily modified to include a Monte Carlo simulation to help analyze risk.

Working through oilfield accident investigations often exposes how different each person approaches troubleshooting. What may seem obvious to one person is not so obvious to another. Never assume someone will choose the optimal option when dealing with an oilfield problem. Furthermore, if you are a manager, do not think that a Company Man will do what you would do to solve a problem. If you want a problem solved in a specific manner, put it in writing. Something as straight forward as calling the base if there is a problem is subject to interpretation.

For example, during the Deepwater Horizon trials, the two Company Men on the rig were indicted by a federal grand jury on 23 counts, including 11 counts of seaman's manslaughter based on a number of findings, including failing to communicate with the engineers onshore that there was a potential problem.[2]

"We look at problems happening halfway across the world and we think, 'Well, that's their problem.' But it's not. . . . When you solve somebody else's problem, you're solving a problem for yourself because our world today is so interconnected."

Queen Rania of Jordan[3]

The indictment alleged that the Company Men on Deepwater Horizon:

*"**Negligently or grossly negligently <u>failed to phone engineers</u>** onshore to advise them during the negative testing of the multiple indications that the well was not secure; failed to adequately account for the abnormal readings during the testing; **accepted a nonsensical explanation for the abnormal readings, <u>again without calling engineers onshore to consult</u>**; eventually decided to stop investigating the abnormal readings any further; and deemed the negative testing a success, which caused displacement of the well to proceed and blowout of the well to later occur."*[4]

Communication is critical. If you are on location troubleshooting an issue, make sure to notify the office, engineering, and/or management. Confirm they know what is happening. Do not assume communication is easy or even occurring. It helps to put a communication roadmap in place using tactical scenario planning methods. One of my preferred methods to address the issue of effective communication, is a flowchart. If something happens in the middle of the night or on the weekend and you want to ensure effective communication, agree with everyone beforehand on the optimal process to notify the team. Then put it in writing to ensure there is no ambiguity on the desired course of action.

Consider employing various tactical scenario planning implementation methods for each critical aspect of oilfield operations. Remember, if there is not a clear straightforward plan in place, different people will do different things when troubleshooting issues on location. That is not necessarily a bad thing because someone may come up with an innovative approach. However, someone may also employ a method that causes a train wreck when it did not have to happen. If, where, or when you employ these methods, they should be customized for the area and operation at hand. Consider three flowchart examples on the following pages.

Communication Plan of Action

The following flowchart illustrates a generic communication plan to help avoid issues due to lack of awareness.

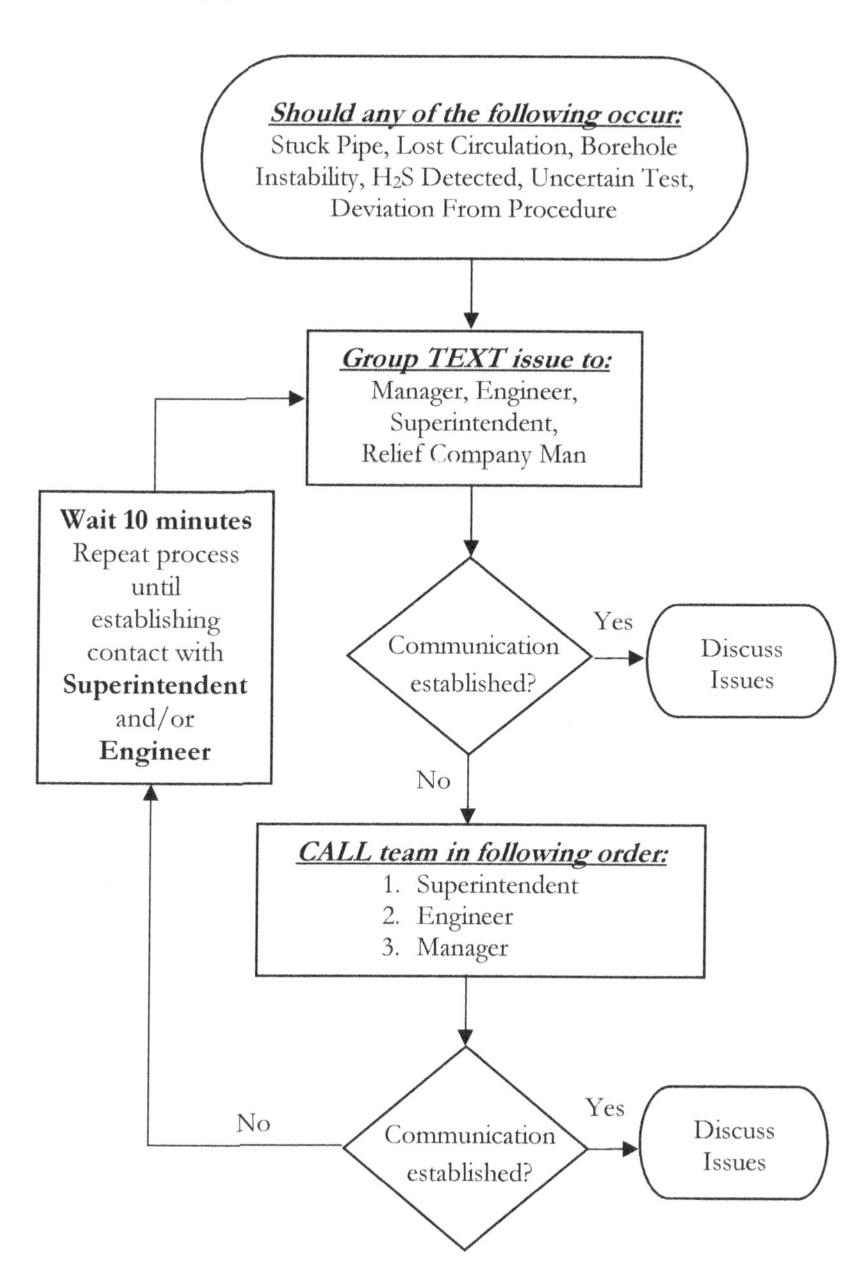

Drilling Lost Circulation Contingency Plan

The following flowchart illustrates a generic lost circulation plan to help the decision making process after successfully cementing 9-5/8" intermediate casing and then drilling the curve on a horizontal well in an area plagued with problems.

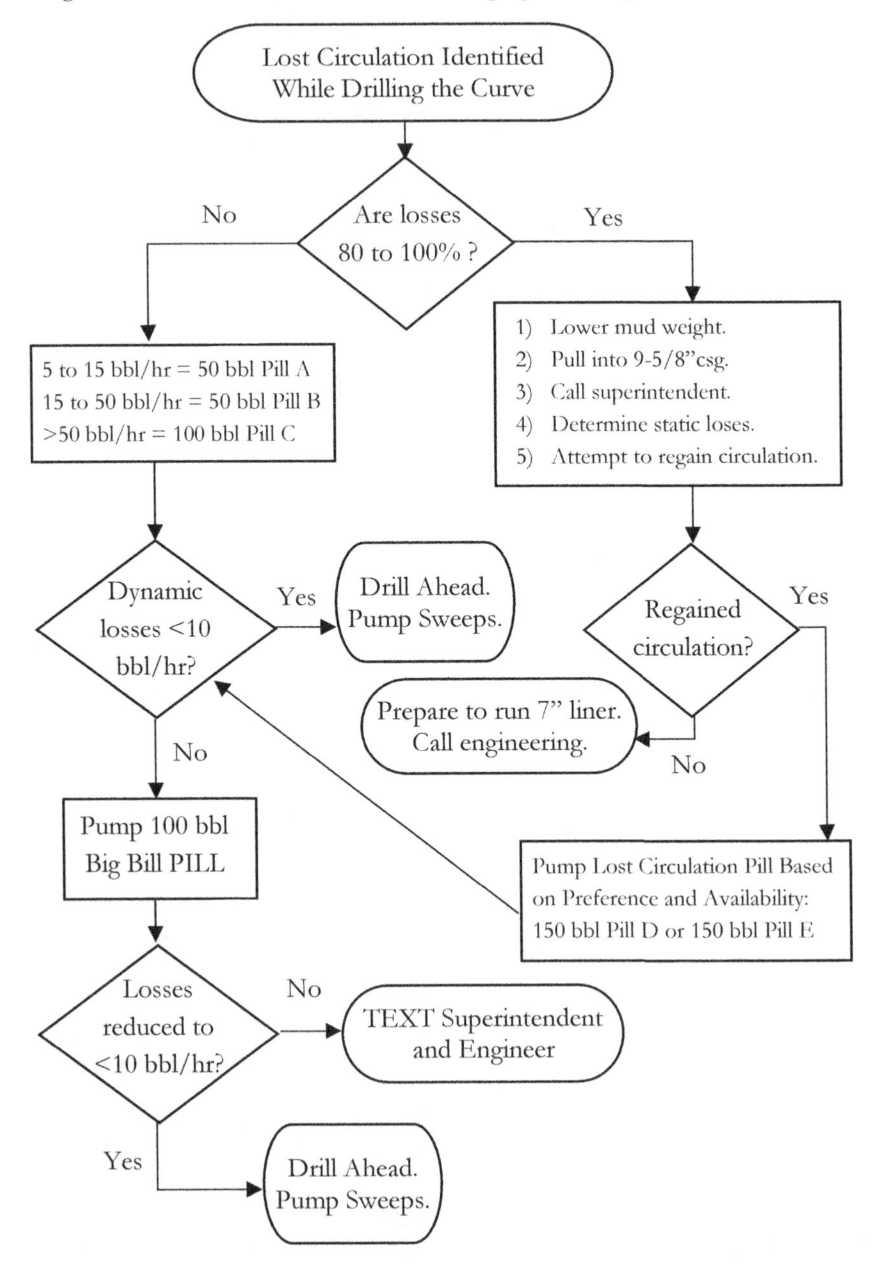

Fracturing Screenout Contingency Plan

The following flowchart illustrates a generic multi-stage horizontal shale plug and perf fracturing screenout contingency plan to help avoid screenouts as well as deal with a screenout once it has occurred.

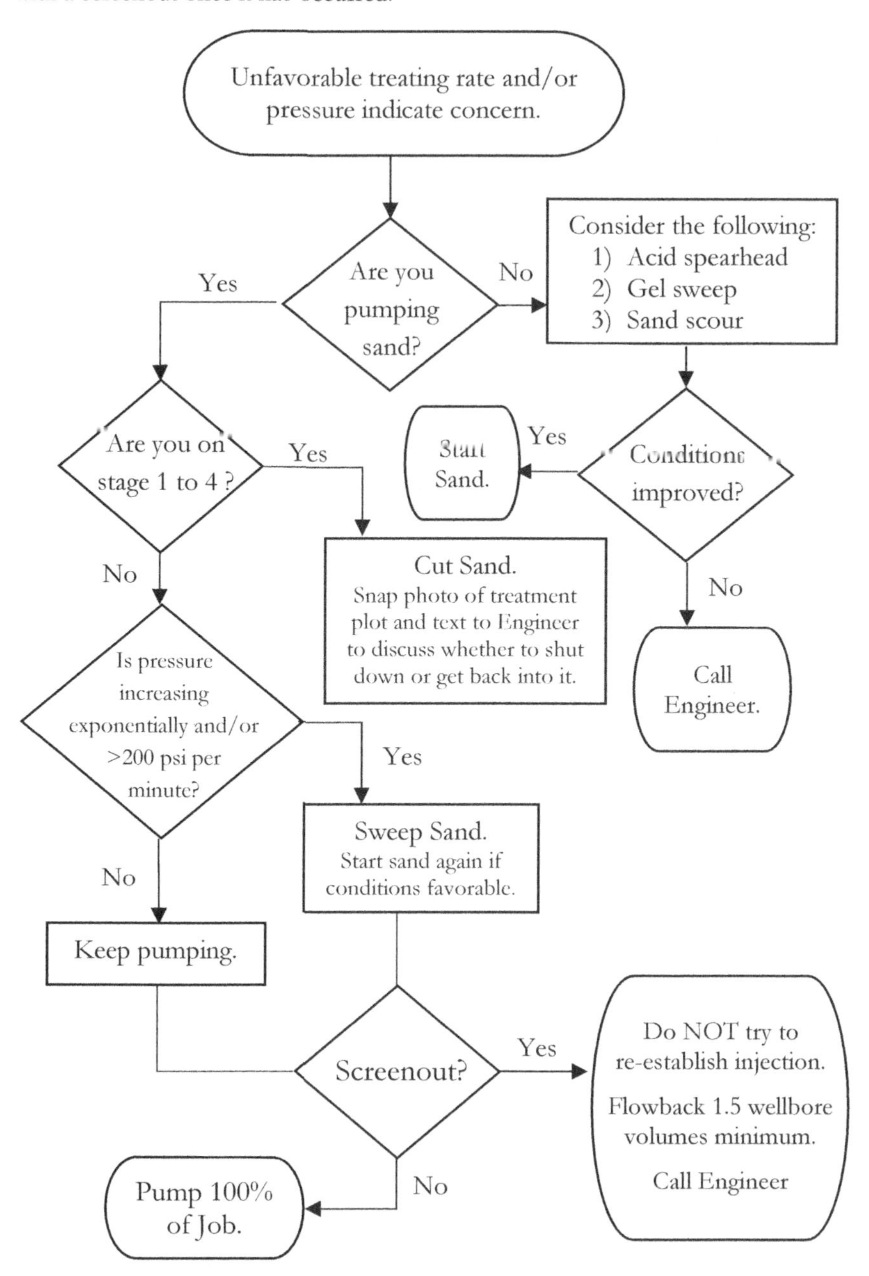

X

CORE TACTICS: 41 - 50

You can improve the entire oil and gas industry, even if it's your first day on the job. Don't believe me? Consider this: The average person meets 10,000 people in their lifetime.[1] If you share valuable oilfield knowledge with 0.10% (10 people) who, in their turn, share the same knowledge with 10 other people, within less than a year the knowledge will spread across the entire industry, due to the multiplier effect. The intelligence "goes viral," spreading quickly, and the industry takes another progressive step towards perfection.

The Internet of the oil and gas industry is its people. A ton of information is shared between hands, especially when everyone is working on location. Since this industry is a tight-knit community, information can travel quickly. The key is effective communication. However, due to the sensitive nature of most oil and gas incidents, and complexity of the situations, detailed information is not easily available. Therefore critical information is not shared. OSG is designed to help change this.

"These new men had never heard about the 'roughneck' who had fallen into the discovery well, so they paid no attention to the unhappy girl who wandered here and there calling his name."

Upton Sinclair, *Oil*[2]

Viral Knowledge In The Oilfield

In January of 2017, oilman Earl Baker Conrad, came across a significant wealth of knowledge shared via emotional oilfield stories. He shares this information with 10 friends in February across multiple locations as he pumps cement, logs wells, and runs tools. He is the embodiment of the ultimate renaissance oilman.

January: Earl B. Conrad acquires knowledge.
February: Shares with 10 friends.
March: They each share with 10 friends. Now 100 folks are safer.
April: 1,000 workers' intelligence increases, spreading at the same rate.
May: 10,000 people learn the valuable information.
June: 100,000 people in summer; some folks go overseas to work.
July: 1,000,000 increase their knowledge across the globe.
August: Fatality rates and accidents start to decline.

If it were only that easy. Unfortunately, the spread of oilfield knowledge is often very slow. It's too much work, too complicated, and boring to most folks. Others hoard information. Why should they share? That's not their job.

I hope you don't feel that way. Safety is everyone's job in this business. If you see something, speak up. Share useful information. One day, you may save your company, your life, or the lives of your teammates and oilfield brethren.

"The companies that are the fittest and that can adapt the fastest to the new industry reality will prevail...and we all know the story of the dinosaurs."
Patrick Schorn, Schlumberger President of Operations[3]

41. ALWAYS: Have a Backup Plan

Every day in the oilfield we make high-dollar, high-risk decisions, without knowing exactly how it is going to turn out. No one knows the future. If you did, you could own the world in a week. Life is uncertain, and the oilfield is not exempt from this fact of life. However, sometimes people forget this and act as if nothing can go wrong. When the stakes are high, operating without a backup plan exposes yourself to the worst possible outcome: "Plan A" fails and there is no "Plan B." Nothing is worse than to be in this situation, backed into a corner and trapped. The only way out is self-destruction: to yourself or your operation.

No Contingency Plan Results in Catastrophe

It was a busy location at the time of the accident. A well was being flowed back at the same time the wellsite facility and tank battery were under construction. At the morning safety meeting, several concerns were addressed about the operation. Visibility was low due to overcast and fog. The location was on the side of a mountain, small, tight, with only one way in or out. Flowback was dealing with a sour well, with H_2S above 500 ppm. They were flaring South of location. In this part of the world during this time of year, wind steadily blows North to South and rarely changes direction. Concern was expressed at the safety meeting that if the wind changed direction, there was no secondary flare system rigged up North of location to which flow could be diverted. The flowback crew explained that their flare ignition system was fail-safe, with multiple backup systems. A few people murmured with concern but ultimately agreed that everything would be okay, or at least did not object aggressively, or try and shut the job down. Operations commenced without issue.

After a large, catered lunch, everyone was feeling a bit sleepy. The wind had changed direction, placing location downstream of the flare. The crew was a little slow to notice and react. A few minutes later, a slug of fluid made its way through the system knocking out the flare. The auto-relight system was having trouble. One of the flowback hands ran to get the flare gun. However, he only had one cartridge, which he fired towards the flare unsuccessfully. Simultaneously, all of the flowback hands were running towards the wellhead, screaming at the facility construction crew to run and get away from the gas that was blowing back on them. One of the flowback hands went to shut the well in. The master valve jammed, and there was no backup wellhead valve. He ran upwind to shut another valve closer to the flare stack when he was hit by the gas and dropped.

At this point, the 8 remaining people on location were backed into a corner with only one option   jump off the side of the mountain. One by one, people started to drop from breathing in the hydrogen sulfide. Two people were able to jump. Only one survived the fall. After around 20 minutes, the auto-ignite system was able to relight the flare. With the gas flaring again, the sole survivor crawled up the side of the mountain and back on location to shut the well in. He did so successfully.

Operations Analysis

Emergency responders were not able to save any of the people that were knocked down by the H_2S gas. Too much time had passed. The accident investigation determined minimal redundancy systems, minimal backup equipment, poor emergency escape planning, and no H_2S contingency planning as contributing factors to the fatal accident. Management from the service providers and the operating company were terminated as a result.

"Every assignment is tough. Even the smallest one can be most dangerous."

Red Adair, Oilfield Firefighter[4]

42. KNOW: The Big Picture

Information and knowledge help prevent train wrecks in this business. Regarding big picture oil and gas finance, strategy, and planning; although understanding corporate level information and decisions is not necessary to execute field operations, it can help you understand dynamics that could impact your operation and your job.

Corporate Strategy Decision Leads to Disaster

A public oil and gas company was entering the 4th quarter and wanted to ramp up production to exceed year end projections. The subsurface team had successfully identified the core area of a new shale play presenting a potential opportunity. The Vice President of Strategy & Planning, along with other executives, made the decision to increase investment in the core of this new shale play by picking up several rigs and moving them into the area. Furthermore, there were a number of horizontal wells drilled and waiting on completion in the core. The executives decided to pick up another frac crew and send it into the core to get all the wells online before the end of the year.

A recently promoted Operations VP, previously manager of reservoir engineering, was an advocate of the plan. The Asset Managers thought the core was not big enough to handle the increase in drilling and fracturing activity simultaneously. However, they were overruled by the new Operations VP who demanded they implement the new plan. The field superintendents were not fans of the new VP because he had insulted them during a meeting. They remained deceptively quiet, in a Machiavellian sort of way. The Superintendents agreed with the Asset Managers, but secretly wanted the new Operations VP to torpedo himself.

The Drilling and Completions Company Men and crews in the field were not informed of the new strategy and ramp up in the core area. They noticed the increase in activity and that the area was getting crowded, but were more focused on their individual wells and daily tasks. Two drilling rigs, 1.5 miles apart, were both drilling the lateral a few days from TD, when a frac crew showed up in the same area. The frac operation and drilling rigs could not see each other because they were on opposite sides of a large hill.

On stage 5 of the frac job, both rig crews thought they were taking a kick and shut the wells in. The frac operation, unaware of what was happening, continued to pump. Once the drilling and frac operations realized that the frac job had fractured into both drilling operations, as they were all in the same reservoir, frac shut its operation down. At this point, the drill string was stuck on both offset wells, resulting in lost BHA's ultimately requiring them to sidetrack. The frac operation was directed to flowback their well to help bleed off pressure on the two drilling operations, as well as shutdown until drilling had finished and the production casing cement had set. The total cost of the disaster was over $10,000,000. The new Operations VP was quietly reassigned to a supply chain management position a few months after the quagmire.

Operations Analysis

The new Operations Vice President's arrogance and failure to listen to his team was the root cause of this accident. Of course, the incident should not have occurred for many additional reasons. However, the strategic decision to overload the core area with activity made it nearly impossible for the field team to safely and successfully execute the new strategic plan. Detailed planning and discussion between field operations and management could have prevented the accident. Executives making strategic decisions in a vacuum and shouting down orders to the field without accepting feedback from the guys on the ground is a recipe for disaster.

"Have the right strategic plan and make sure you have stress-tested it."
John Hess, CEO Hess Corporation[5]

43. KEEP: Accurate Detailed Records

Record keeping is a key aspect of our job in the oilfield. Due to the capital intensity of this business, detailed records are critical to ensure accurate information is available to make future decisions, minimizing risk and financial exposure. Making investment and operations decisions with incomplete or inaccurate information can be catastrophic in the oilfield. Remember, before most operations occur, an investment decision is made.

In regards to specific well operations, everyone involved in an operation should have access to the well history data. Reading the history of a well should be a foundational step during the engineering design process and procedure construction. Not knowing the detailed history of a well is setting yourself up to have an undesirable situation or surprise, usually unpleasant.

Poor Record Keeping Results In Hardline Crossing

The directional survey, daily reports, and casing tally were incomplete, indicating total depth was 100 feet before the perforation hardline for a new horizontal well. Relying on this incorrect information, a sliding sleeve system was run across the toe hardline without knowing it.

The land and regulatory department was preparing to file paperwork on the well, when they found the issue after obtaining the final directional survey from the Drilling Engineer. He had forgotten to put the final survey in the well file or cross-reference it with the daily reports and casing tally to confirm they were correct. The well had been on production for a month before the issue was found.

Vertical Well Lost Due To Missing Files

The plan was to clean out a vertical well for recompletion. The decision was to use coil tubing for the cleanout since the well had pressure. Coil had trouble getting to bottom so they picked up a bit to drill out what was thought to be hard paraffin. When metal shavings were circulated to surface, the team became concerned. However, they were making progress and continued to run to bottom where they became trapped.

After spending several days trying to get free, one of the geology techs found a word document buried on the company server, which enlightened the team. Apparently, the well had a casing patch and a complex fish. The metal shavings were from drilling through the casing patch, which had a smaller ID than the casing. Coil tubing got stuck at total depth because there was a fish in the well left from the casing patch operation, which most likely trapped the coil BHA. With all of this "new" information, the decision was made to abandon the operation, cut coil, and plug the well.

Acquisition Based On Inaccurate Well Records

A private equity backed management team was looking to acquire stripper wells with recompletion potential. A large number of legacy vertical wells were purchased from an oil company with the plan to kick off a recompletion campaign to justify the purchase price and generate value for the investors. After employing the strategy on several wells, it was discovered that recompleting these wells was going to cost more than initially estimated and used in the valuation model, significantly changing the economics. The mistake was due to inaccurate well records that did not indicate the correct casing specs used on the majority of wells in the field.

44. MANAGE: Costs and Invoicing

Before generating an invoice, or signing anything on location, confirm all line items with the original bid price. If you are a Company Man or Field Engineer always ask management for a copy of the most recent bid or quote so you can compare it to the field invoice. Never sign a ticket unless you have documentation on what the total cost and all line item costs should be based on the pre-job estimate.

It is fairly common for field invoices to not be aligned with the latest agreed-upon bid price. A mistake in the field ticket can happen for a number of reasons. For example, the service company may not have sent the latest prices to the field. Additionally, the field service leader may forget to adjust the on-location ticket template from a previous customer. You must double check every line item and the total cost with the latest pre-job estimated cost or bid price before you sign the ticket.

Frac tickets require special attention because they can be fairly complex and mistakes are common. Consider building an excel spreadsheet that helps confirm the field ticket. This is something the team can use to confirm the tickets are correct on location. Some operators will not sign frac tickets or other +$1,000,000 tickets on location. One option is to have the ticket emailed into the office, double checked, signed in the office, and a scanned copy sent back to location.

Incorrect Invoices Signed Off By Company Man

Over the course of three months, a Company Man, who worked for an oil company notorious for not paying attention to invoices, performed seven frac jobs, signing the final ticket for each job. The Engineers did not send him the bid price with line item detail and the Company Man did not ask for it. He was only told that if the frac ticket is less than $2.5 million, sign it.

The signed field tickets came in with the final bill and were paid by the oil company. Mid-year, the accounting department decided to audit all frac invoices. Every invoice this Company Man signed was over $200,000 high compared to what it should have been, a mistake of over $1.4 million dollars in total.

Incident Analysis

This incident changed how the oil company approved invoices. The new process included a number of verification steps. First, all service work was bid, regardless of the size, to get a sense of the market and obtain an estimated bid cost. Second, all estimated costs were sent to the field team which constructed a folder and price menu with all recent prices. Third, before signing any invoice, field personnel would refer to the bid estimate. Forth, anyone who signed was required to stamp the invoice and fill out several additional data points including the bid document referenced, which confirmed the line items were correct. Fifth, all vendors were required to attach the bid documents to the invoice before sending it to the office for payment. Once received in the office, an additional process was followed by the engineers and accounting department to double check the invoice.

Lack Of Cost Awareness Results In Lost Acreage

An operating company was in the process of ramping up rigs, doubling the rig count. The operations team was barely able to keep up with the increased activity. The focus was on getting wells safely drilled and completed as quickly as possible. Cost was being accounted for but it was not tracked with an extreme level of detail. Based on the daily reports, the field estimated cost appeared to be below the AFE estimates. The team was celebrating its ability to drill wells under budget and about to reduce future AFE's, when the actual costs started coming in. To the surprise of the entire asset division, especially management, actual costs were 45% higher than the field estimated cost in the daily reports.

Incident Analysis

Many costs were not captured by the company men in the field and these omissions were not identified by the engineers. Very quickly the asset division exceeded their budget. All operations were then shut down by the CEO. As a result, expiring acreage was not drilled and; therefore, lost to a savvy new startup oil and gas company patiently waiting to grab up the expired acreage at a discount.

This incident changed the field cost documentation process. Previously the oil company depended primarily on the company men to document all costs in the daily report after the tickets were signed on location. The new process included a projected daily cost estimate for each day of the operation. This number was then compared to the field estimated cost at the end of each work day. Each day at 5 p.m., the company men and engineers were required to review total field estimated cost for the day to ensure nothing was overlooked or left out of the daily report.

45. INVEST: In a Drilling Location Inventory

The root cause for undesirable oilfield situations can be found not only on location, with operations activities, but also in the office, with decisions made regarding resource development, strategy, and planning. One area often overlooked is the inventory of locations ready for field operations. A location ready for active oilfield operations may include locations with:

1. Detailed predrill subsurface research performed.
2. Analysis indicating the well will meet type curve projections.
3. Economics exceeding hurdle rate or expectations.
4. Approval for drilling by management and partners.
5. Procedures thoroughly reviewed by field operations.
6. Location prepped for rig or frac mobilization.
7. Operations logistics arranged and scheduled.

Running an active business unit with multiple rigs and frac crews, without locations ready for operations, puts a lot of pressure on the entire team to have a place for the rig or frac crew to move to. For example, when a drilling rig or frac crew is getting close to finishing on the current location, knowing the plan for the next location helps ensure success, especially with field planning, safety, efficiency, and cost control. Not having a plan or waiting until the last minute can set an operation up to fail before it begins. The situation can be exacerbated when drilling activity is increased before building an inventory of locations for the additional rigs.

Increased Rigs + Minimal Inventory = Disaster

An operating company hired a new CEO with a finance background to lead a corporate turnaround. He directed his newly installed finance team to build an advanced model to determine the optimal strategy to create value for shareholders. The model suggested cutting all the rigs across the company, except for one area in Texas. In that area, the model suggested to triple the rig count from 2 rigs to 6 rigs. The new CEO praised the small Texas asset team as the area generating the highest returns. The team working the Texas asset was proud that they had finally received recognition for their hard work.

For the past two years, this team carefully selected the best drilling locations and worked meticulously on cost control. It was a strong team that worked well together, especially field personnel and the office technical staff. There were many friendships within this group that helped ensure communication was effective. They rarely had a train wreck type situation.

The new CEO replaced the Texas asset VP with a new ambitious VP who would lead the increase in drilling rigs. Arrangements were made to immediately mobilize rigs from areas of the company that were cutting rigs. With only a few weeks to prepare, the Texas asset team was in a panic to prep locations for the new rigs. They were not able to follow their previous location selection process of meticulous analysis, internal debate, and approval because there was no time.

The six rigs started burning through the small inventory of locations they had previously built for two rigs. With no inventory, whoever proposed a location, if it sounded good, they drilled it. They drilled anything that was ready. The geology, drilling, and completions staff could not generate the geology prognosis and operations procedures fast enough. Previously, the procedures were tailored to each area of the play, now it was cookie cutter. To help out, personnel were moved into the Texas asset from other areas. Unfortunately, most of these folks blindly applied designs and methods that they were previously using in other areas of the company to the Texas asset, and it did not work.

Each week there was a new train wreck driving costs into the stratosphere. The new VP did not understand what was happening. Then the well results started underperforming expectations. At first, production was slightly below type curve, then the type curves had to be reduced. Month after month the curves were dropped until the asset was deemed unprofitable. With the asset in disarray, the new VP almost had a nervous breakdown during one of the management meetings. Based on everything that was happening, the CEO decided it was time to sell the Texas asset.

Incident Analysis

Management did not give the asset team enough time to build an inventory of good locations before mobilizing the additional rigs. Due to this critical mistake, the CEO's plan backfired, destroying value. A savvy private equity backed management team was following the developments and was able to acquire the Texas asset at a bargain price. The small management team was able to turn the asset around and sell it to another large oil company after only a few years. The small management team became millionaires as a result.

"People know. They want to follow a leader. If you're not one, they are not going to hire on. They are not going to follow you."

Harold Hamm, Billionaire Oilman[7]

46. NEVER: Depend on One Valuation Model

Redundancy allows for successful longevity in this business. Similar to running two tested barriers, when performing oil and gas valuations you can avoid costly mistakes by having two people with two different economic models perform the same valuation and compare results. Many errors will be caught by doing this.

If you can't have two people use two different models, at least have one person use two different models. Running an excel model and comparing to a standard oil and gas software based valuation is a good way to catch errors. Both tools have the ability to be significantly incorrect, and sometimes, it can be very difficult to find the errors. Depending on the size of the asset under evaluation, errors can easily be in the tens of millions to the hundreds of millions of dollars.

Acquisition Flawed By $20MM Calculation Error

A small private oil company was selling their acreage position, including 12 producing horizontal shale wells. The acreage position was derisked and had a dependable type curve. Since it was a desirable position, several interested parties were in the process of making offers. For an interested mid-sized independent operator, a Reservoir Engineer was tasked with performing the valuation. He chose to use a well-known economic evaluation software, to which his company had subscribed, to perform the valuation. Based on his analysis, the company offered $100MM for the asset and ended up winning.

After the deal closed, another Reservoir Engineer was tasked with running a sensitivity analysis to optimize the development plan. This Reservoir Engineer decided to build the model in excel. He ran into problems when he could not match the initial valuation. After spending a day comparing the two models, the issue was found. The original valuation had several small errors including a decimal being off one place. With the corrections made, both models suggested a valuation $20MM lower than what the company paid.

The CEO went into a rage against the VP, threatening him with physical violence. A meeting was called where the CEO and other executives berated the reservoir engineering department and A&D team. After about four hours, just when everyone was at the breaking point, a Completion Engineer and Drilling Engineer stood up and suggested a plan to increase the value of the asset. They proposed increasing the lateral length from 4,500 feet to 10,000 feet, combined with reduced cluster spacing and increased proppant loading. Based on initial analysis, they thought by employing these methods they could improve the economics which would more than justify the purchase price. The CEO approved several trials of the plan, which ended up working. With the new design, the value of the asset increased from $80MM to $200MM.

Incident Analysis

No economic model is perfect. Each has errors. The key to success is to minimize the impact of the imperfections on the end result. As this incident has shown, a decimal in the wrong place can have a significant impact on multimillion dollar acquisition decisions. This is why it is critical to double check valuation models using two independent models from two different people. Fortunately, the operations team stepped up with an idea to create value by employing a new engineering design.

"It's not a spreadsheet. Business is people at the end of the day. And keeping them motivated and excited and pointed in the right direction is the secret to business. But it's not the math, it's really the team."

Clay Williams, CEO National Oilwell Varco[8]

47. NEVER: Solely Utilize Point Forward Economics

Operations often do not go exactly as planned. As a result, decisions and adjustments are made to ensure success. Safety should always be the top priority. Once safety is addressed, cost and economics are often taken into the decision making process. For example, in certain situations when facing adversity, the decision may be to skid the rig due to a fish, call TD early when in the lateral due to hole problems, not complete a well if the reservoir is determined to be undesirable, end (flush) a frac job early, skip several frac stages, stop completion operations entirely, or plug and abandon.

Since each decision may have a significant impact on profitability, in terms of immediate cost and production, point-forward economics is often employed to quantify the impact of a course of action. Point-forward economics do not account for past (sunk) cost. The methodology assumes it never happened, and only looks forward, to future estimated cost and future potential production performance. Used incorrectly, point-forward economics can and will bust your budget, destroy value, and potentially bankrupt your company. This is best illustrated with a story.

Reservoir Engineers Cause Financial Fatality

Xcellus Oil was drilling a horizontal shale well, Mahoney Pitt #4H, estimated to take 20 days spud to TD. The team had drilled 4,500 ft of lateral by day 45 with 3,000 ft of lateral remaining, based on a planned 7,500 ft lateral length. The well was plagued with many problems. At the time, they were making little progress, only a few feet per day. Mahoney Pitt #4H had lost circulation issues combined with several BHA failures. The well was also sporadically flowing high rates of natural gas.

The Drilling Engineer, Drilling Manager, and Field Crew suggested calling TD early, attempt to regain full returns, and then run production casing. Field estimated well cost was currently at $13,000,000; the AFE was for $8,000,000. The Vice President requested the Reservoir Manager run economics and provide a recommendation. The Reservoir Manger, did not ask questions, he did what he was told and had the reservoir engineers run point-forward economics. They quickly concluded that spending an additional $2.0 million dollars to reach TD generated over 100% ROR for the additional 3,000 feet of lateral. The VP estimated that it would cost an additional $2.0 million to reach TD. Where he got that number from nobody knew. The decision was made to keep drilling.

After another week, they made an additional 500 feet. However, field estimated cost was now at $15,000,000. They burned through a significant amount of money, losing high dollar oil-based mud. The team reconvened and the VP requested reservoir engineering run economics. They did what they were told and utilized point-forward economics again, assuming all past costs were sunk cost. The economics suggested to keep drilling. At this point, everyone was optimistic that the difficult drilling problems would soon be solved and that they would reach TD within a few days.

A holiday weekend was coming up and much of the team was going to take vacation. Another week went by. At the Tuesday morning meeting after the holiday, the team found out no progress was made and that they were stuck. Field estimated cost was now at $20,000,000. Apparently, the Company Man realized they had omitted several big items from the daily reports and made the correction over the weekend. At this point, the VP was sweating bullets. He considered asking reservoir engineering to run economics again, when he realized they would always conclude to keep drilling. He realized the methodology was flawed.

Once Mahoney Pitt #4H exceeded $20,000,000 it caught the attention of the CEO and COO who were furious. The COO called the Drilling Manager who explained that the VP and Reservoir

Manager were calling the shots. An emergency meeting was called where the COO told the VP to call TD and end the train wreck.

Incident Analysis

The COO explained that the VP and reservoir engineers fell into a value trap with point forward economics. When historic costs are not taken into consideration, the reality of the operational situation not properly understood, a risking factor not applied, and no hard stops established, the decision is almost always to keep moving forward with operations. To keep dumping capital into a money pit. Sunk costs continue to accumulate until you run out of money or you are fired. Additionally, the more money you dump into a well, the more emotionally attached you become. To make sure they learned the lesson, the COO eliminated the annual bonus from the VP and Reservoir Manager.

48. NEVER: Underestimate the Downside Risk

Humans are by nature optimistic creatures. That is a great thing; it is productive to be optimistic. We should be optimistic. In my opinion, the oil and gas business attracts the most optimistic people. You have to be optimistic to be in this business because most exploration wells are economic failures. Shale has changed that to a certain extent. However, the commodity price risk still exists, now more than ever. As I write this, the industry is recovering from one of the biggest commodity price downturns in decades, forcing many companies into bankruptcy.

When working hard to make something successful, if things do not work out, it can be difficult to accept. Investing money, time, and hard work into a project emotionally attaches you to that project and to its success. I am a firm believer that hard work pays off. However, sometimes it does not pay off as soon as you would like.

In oil and gas field operations, when working hard to make a job successful on location, it can be difficult to accept if the job fails, or something that you worked on all day or all month does not turn out as you would like. For example, a lot of time, money, and hard work goes into taking a well to the point that production casing cement is pumped. When performing testing to confirm success, whether with a negative pressure test or cement bond log (CBL), everyone involved in the operation wants to see no pressure or a successful interpretation of the CBL. When things are not clear on the CBL or negative test, it can be difficult to accept. To confirm results, often follow up testing is performed, and it should be. Always confirm and double check. However, don't explain uncertainty away. Being invested, in terms of hard work, time, money, reputation, and emotion, can influence you to view an interpretation unrealistically, be too optimistic, and/or underestimate the downside risk. Many things in this business are subjective to a certain extent. Unbiased interpretation and analysis is required. As much as everyone involved in the operation, from the office executive to the field supervisor, wants and hopes for no issues, accept reality for what it is. Identifying failure is part of being successful.

Accept reality for what it is and then address it. When identifying potential problems, there are opinions and there are facts. As strong as someone's opinion may be, do not confuse it with the facts. Everyone wants an operation to be successful. Therefore, it can be difficult to unbiasedly assess risk. No one is immune from bias, emotion, and over confidence, impacting their ability to objectively assess risk. Not even a CEO with 35 years of experience, as the following story illustrates.

CEO Decapitated During Stimulation Treatment

Bob Goodman, CEO of Onetime Resources, had 35 years of experience, 20 of those years were in the field. He was a legend in the oil business and well respected by all, especially field crews. He had years of experience with solving problems and helping people get out of train wreck situations. Onetime Resources was considered by many to be the most safety-focused company in the oilfield. Bob was credited with this achievement, as Onetime Resources had the top safety record among all independent oil and gas operators ever since Bob became CEO ten years ago. Bob held safety above everything else and instilled a corporate culture of safety. Bob visited the field frequently to ensure everyone was following company rules and running a safe operation. Bob always put safety first, except for one time, and one time was all it took to lose his life in a tragic accident.

Bob was visiting the field for an executive level safety audit. The group decided to visit a frac job, as Onetime Resources was performing multistage fracturing operations on a horizontal shale well. The well was over budget due to issues during drilling. Onetime Resources was also under financial pressure due to the oil price collapse, which put additional stress on Bob and his executive team.

The executives were monitoring the job from the frac van when a leak developed on one of the frac iron risers connected to the goat head on the frac stack. The Line Boss radioed to the Treater to shut the pumps down. The Treater was about to shut down when Bob told him to hold on.

Line Boss: *"We have a leak, need to shut down."*

Treater: *"Okay, lets…" (Bob interrupts him mid-sentence)*

Bob: *"Hold on. It does not look that bad. Were almost done."*

Company Man: *"Bob, as you know company policy is that we immediately shutdown, and/or cut sand and drop rate."*

Bob: *"Yes, I know but I really don't want you to have to get coil out here if we can't get back into it. Let's try to get 100% of this sand away."*

Company Man: *"Lets at least cut sand and drop rate to prepare to shut down if something happens."*

Bob: *"Let me get a closer look at the leak."*

Treater: *"Service company policy, you can't go in the hazard zone around the wellhead sir."*

Other Executive: *"Bob, this really isn't like you. Don't you think we should shut down?"*

Bob: *"I'll be right back."*

Bob left the frac van and walked towards the wellhead to get a closer look. He moved in a little closer to assess the leak when the iron parted at a connection. One of the risers swung around striking Bob in the neck. The crew immediately shut down. Bob was found decapitated next to the wellhead. The safest man in the oilfield was killed due to a lapse in judgement. It only took one time, just one mistake, for the safest man in the oilfield to lose his life.

49. ALWAYS: Be Respectful

A professional "Tier One" oilfield operator never shows disrespect to others, regardless of experience or position. Teasing, bullying, or insulting others can be a distraction and contribute to or cause an accident. Additionally, if you are consistently disrespecting others and clowning around, you will not be taken seriously by your colleagues or management. There is a fine line between having fun and insulting someone. It is hard to know where that line is because everyone is different. It is not worth damaging your career in the oilfield. If you are making fun of someone and they decide to go to the human resources department, you could lose your job or get reprimanded. It is just not worth it.

Unfortunately, in certain organizations, management can also be disrespectful. This can promote a hostile and toxic work environment which will almost certainly contribute to undesirable oilfield situations. Due to the amount of risk in the oilfield, a company cannot afford for its employees to be distracted due to management disrespecting subordinates. There are already enough problems to deal with.

My experience with this issue, particularly with vendors, new managers, new employees, minorities, and women, over many years in the United States and globally, suggest it is a problem, especially on location. Frequently, I have seen engineers, company men, frac treaters, and drillers disrespect roughnecks, frac hands, salesmen, and other service company personnel.

Service providers work hard to add value for operating companies. When company men insult vendors, it does not help the relationship and can end in retribution. If drillers are constantly disrespecting their crew on the rig, it can lead to an oilfield incident. It is not unheard of for someone to cause an incident to occur as revenge to the frac treater, driller, or manager that has been yelling at them and disrespecting them all day. In the end, the operating company typically pays the bill when retribution is performed to the well or operation.

For example, in a restaurant, it is not advisable to insult the people who are handling your food. This is a well-known and followed rule because if you insult the servers, they can do something undesirable to your food. It is similar in the oilfield. If you insult field personnel, they may reciprocate in one form or another, and in many cases, the operating company will not even know that their multi-million dollar train wreck was the product of payback for disrespect.

If you work for someone that is constantly yelling and disrespecting you, I advise that you find another job. On the operating side of the business, if you witness a hostile service provider environment on location, find a new vendor or crew to perform your work. For example, a driller or toolpusher who is constantly yelling, disrespecting, and bullying the crew probably has a crew that despises the toolpusher or driller. If something were to happen to the operation to get that toolpusher or driller fired, the crew may allow it or cause it to occur. They may lose their job and walk away. However, the oil company is left with a multi-million dollar train wreck to deal with.

SURVIVAL SKILLS

Avoid participating in oilfield pranks. They tend to spiral out of control as people try to outdo each other until an altercation occurs, usually with fists.

Geophysicist Verbally Stressed To Death

The importance of professionalism is best illustrated by studying a court case in which lawyers use a disrespectful manager as evidence of undesirable work conditions which contributed to the death of a geophysicist: an oilfield subsurface expert often focused on seismic and geophysical related analysis.

Key excerpts (abridged and adjusted) from legal documents are summarized on the following pages:

"Ronald Rush, owner of Big Rush Oil, offered Mr. Friendly a position as a geophysicist for more money than he was currently making in addition to overriding royalties in all oil, gas, and mineral leases acquired. While on a business trip, Mr. Friendly was running to board the company plane and had stopped to catch his breath. Mr. Rush was rushing him and telling him to get on the plane. Mr. Friendly began running again when a severe angina attack [chest pain] hit him. That same day he was admitted to the hospital. Mr. Friendly suffered from atherosclerotic occlusive heart disease which consisted of 95% blockage in his right coronary artery and approximately 100% blockage in his left coronary artery. A team of physicians performed open heart surgery; however, Mr. Friendly died in the recovery room of massive heart failure.

The trial court determined that Mr. Friendly's employment was stressful. He had to find oil for his company, a task that required physical as well as mental exertion. Based on his suggestions, the pursuit to find oil at a particular site would become an expensive gamble for his company. The stress from these activities was compounded unnecessarily by the conduct of his employer.

It is apparent from the record that Mr. Rush was of a temperamental nature. The memorandums he wrote to his employees show that he wanted them to fear him. In one memorandum addressed to landmen, geologists, geophysicists, engineers or To Whom It May Concern, he concluded it by saying:

> *'Do not speak to me when you see me. If I want to speak to you, I will do so. I want to save my throat. I don't want to ruin it by saying hello to all of you sons-of-bitches.'*

In a memorandum addressed to all monthly salaried personnel, he commented:

'There is one thing that differentiates me from my employees. I am a known son-of-a bitch, and I care to remain that way.'

He would call Mr. Friendly at home at 1:00 A.M. and other odd hours to have lengthy discussions with him. If Mr. Friendly disagreed with him, Mr. Rush would threaten to fire him. A few months before Mr. Friendly's death, Mr. Rush began threatening to fire him if he would not renegotiate the employment contract leaving out the overriding royalty clause. Mr. Friendly would not agree to give up his overriding royalty but Mr. Rush continued to pressure him to do so.

The employees even talked about the stress Mr. Friendly was enduring because it was so apparent. Mr. Friendly continued to be under pressure from Mr. Rush until his death. Mr. Friendly did not want to submit to surgery. An employee of Big Rush Oil, testified that Mr. Rush ordered Mr. Friendly to have the surgery or he would be fired. Mr. Friendly was subject to angina pectoris (chest pain) attacks. In the last few months before his death, the attacks became more frequent and correlated with the extra stress he was subjected to by his employment. Mrs. Friendly testified that the phone calls during the middle of the night from Mr. Rush as well as the threats to fire her husband were upsetting and causing great stress upon her husband."[10]

Incident Analysis

Oil and gas is a tough and stressful industry. On top of managing risk in this business, each person has personal things going on in their lives that they have to deal with. Giving people a hard time on top of everything else is just not right, not safe, and not smart. Regardless of what is happening at work, always be nice to people. In this particular case, Mr. Friendly was verbally abused on a regular basis. As a result, the court determined that his death was related to his employment.

"Be kind, for everyone you meet is fighting a harder battle."

Plato, Philosopher[11]

50. TAKE: Control of Your Oilfield Safety and Success

You are in control. This point is so important that it warrants repeating. You are in control of your safety and success in this business of oil and risk. Not only are you in control, you are responsible and obligated to protect yourself, protect your colleagues, protect your company, and protect the great business of Earth exploration. If you have read this entire book, I hope you have embraced that anything can go wrong and it can happen at the worst possible moment. I also hope you have embraced that everything is preventable and that you can prevent it if you take action.

Ask yourself why you work in the oilfield, why do you do this job of all possible jobs, why this one? Is it for the money, is it for the comradery, is it for adventure, is it for your family? Whatever you are most passionate about, bring that passion to work every day. When it comes to safety, bring that passion because it will allow you to be successful for a long time.

Honor and respect the thousands of people who gave their lives in this business. Their stories provided you with valuable lifesaving lessons. Remember them when it matters most, in the middle of an operation, when everything is going fast, and the last thing on people's minds is disaster. No one thinks it's going to happen to them. No one wakes up in the morning and thinks, today is my last day, I will not be coming home from the field.

Safety, environment, then the job. Never hesitate to shut down a job for safety. You will not be penalized by your company or a customer company. It can be hard to know when to take action. When under stress, in a panic, it is hard to remember everything. Combine that with lack of attention to detail, and it provides the setting for an undesirable situation. Hopefully, I have helped make some of these stories memorable to you, so they stick in your mind when it matters most. When you are under pressure and out of time.

OILFIELD SURVIVAL GUIDE

SOURCES

Sources

Other than Piper Alpha and Deepwater Horizon, the names of incidents, oil and gas reservoirs, companies, and people involved in the situations contained within this book, have been changed to protect the privacy of businesses and individuals. Any resemblance to actual persons, locations, or companies is purely coincidental.

Preface

1. Titus Maccius Plautus, *Persa*, Act IV, (c. 250 – 184 BC)
2. Translation by Bonnell Thornton, 1767

Chapter One

1. Bellmont Research Team, *Mr. Buffett, What are More of Your Secrets?*, (Apr. 11, 2011), *available at* www.bellmontsecurities.com.au/mr-buffett-what-are-more-of-your-secrets.
2. Charles T. Munger, *Poor Charlie's Almanack: The Wit and Wisdom of Charles T. Munger*, (2005), *available at* www.goodreads.com/work/quotes /3406894-poor-charlie-s-almanack-the-wit-and-wisdom-of-charles-t-munger
3. Marsh, *The 100 Largest Losses 1974-2013*, 23rd Ed, (2014)
4. T. Boone Pickens, *The First Billion Is the Hardest*, Crown Business, (2009)
5. Wikiquote, *Talk: Abraham Lincoln*, *available at* en.wikiquote.org/wiki/ Talk:Abraham_Lincoln #Did_Lincoln_say.2C_.22My_best_friend....22
6. Federal Aviation Administration, *Northwest Airlines Flight 255, McDonnell Douglas DC-9, N312RC*, Detroit Michigan, (August 16, 1987)
7. National Transportation Safety Board, *Aircraft Accident Report, Northwest Airlines, Inc. McDonnell Douglas DC 9-82 N312RC*, (May 10, 1988)
8. U.S.DOT, FAA, *The Use and Design of Flightcrew Checklists and Manuals*, by John W. Turner, EG&G Dynatrend, M. Stephan Huntley, Jr. U.S. DOT, John A. Volpe, National Transportation Systems Center (April 1991)
9. See *supra*, note 8.
10. National Safety Council, *Workplace Injuries by the Numbers*, Injury Facts *available at* www.nsc.org/JSEWorkplaceDocuments/Infographic-Injuries-bythe-Numbers.pdf (2016)
11. Fortune, *Inside the Mind of Jack Welch*, by Stratford P. Sherman and Cynthia Hutton (March 27, 1989)
12. The American Oil & Gas Reporter, *Companies Must Prioritize Health, Safety Aspect*, by Corinne Westeman, Arkansas Independent Producers & Royalty Owners, Safety Beyond Compliance panel (November 2015)

13. National Aeronautics and Space Administration, NASA History Office, *Explosion of the Space Shuttle Challenger Address to the Nation*, by President Ronald W. Reagan, (January 28, 1986)
14. Foundation for Economic Education, *John D. Rockefeller and the Oil Industry*, Burton W. Folsom, (October 1, 1988)

Chapter Two

1. Gerald Lynch, *Roughnecks, Drillers, and Tool Pushers*, University of Texas Press, (November 1987)
2. Marsh, *The 100 Largest Losses 1974-2013*, 23rd Edition, (2014)
3. National Safety Council, *Workplace Injuries Infographic*, available at www.nsc.org/Measure/Pages/Workplace-Injuries-Infographic.aspx
4. Stanford University, *Commencement Address*, by Steve Jobs, *available at* news.stanford.edu/2005/06/14/jobs-061505/ (June 12, 2005)
5. United States Department of Labor, Occupational Safety and Health Administration, *Safety and Health Topics – Oil and Gas Extraction*, *available at* www.osha.gov/SLTC/oilgaswelldrilling
6. ConocoPhillips, *Heath, Safety and Environment Policy*, by Ryan Lance *available at* www.conocophillips.com/sustainable-development/our-approach/ Documents/Health,%20Safety%20and%20Environment%20Policy_FIN AL.pdf
7. Politico, *Statement Prepared for the Hearing on "Drilling Down on America's Energy Future: Safety, Security and Clean Energy"*, Energy and Environment Subcommittee of the House Energy and Commerce Committee, by John S. Watson a*vailable at* www.politico.com/pdf/PPM136_watson_statement. pdf (June15, 2010)
8. Daniel Coenn, *Confucius: His Words*, BookRix, (June 8, 2014)
9. United States Coast Guard, *On Scene Coordinator Report Deepwater Horizon Oil Spill*, (September 2011) *available at* www.uscg.mil/foia/docs/dwh /fosc_dwh_report.pdf
10. The New York Times, *Gulf Spill Is the Largest of Its Kind, Scientists Say*, by Campbell Robertson and Clifford Krauss (August 2, 2010)
11. The Wall Street Journal, *BP Agrees to Pay $18.7 Billion to Settle Deepwater Horizon Oil Spill Claims*, By Daniel Gilbert and Sarah Kent (July 2, 2015)
12. Fox News, *BP to pay more than $20 billion in record oil spill settlement*, (October 5, 2015) *available at* www.foxnews.com/us/2015/10/05 /bp-to- pay-more-than-20-billion- in-gulf-oil-spill-case-largest-settlement-in-us.html
13. BP, *Gulf of Mexico restoration*, (October 2016) *available at* www.bp.com/

en_us /bp-us/commitment-to-the-gulf-of-mexico/gulf-mexico-restoration.html; United States Department of Justice, *U.S. and Five Gulf States Reach Historic Settlement with BP to Resolve Civil Lawsuit Over Deepwater Horizon Oil Spill, Fact Sheet* (October 5, 2015) *available at* www.justice.gov/opa/file/780696/ download; BP, *BP estimates all remaining material Deepwater Horizon liabilities*, (July 14, 2016) *available at* www.bp.com/en/global/corporate/press/press-releases/bp-estimates-all-remaining-material-deepwater-horizon-liabilitie.html; The New York Times, *Judge Accepts BP's $4 Billion Criminal Settlement Over Gulf Oil Spill*, by Clifford Krauss (January 29, 203); Reuters, *BP gets $4 billion from Anadarko for oil spill costs*, by Tom Bergin and Moira Herbst (October 17, 2011); United States Department of Justice, *Transocean Agrees to Plead Guilty to Environmental Crime and Enter Civil Settlement to Resolve U.S. Clean Water Act Penalty Claims from Deepwater Horizon Incident,* (January 3, 2013) *available at* www.justice.gov/opa/pr/transocean-agrees-plead-guilty-environmental-crime-and-enter-civil-settlement-resolve-us; The New York Times, *Halliburton to Pay $1.1 Billion to Settle Damages in Gulf of Mexico Oil Spill*, by Clifford Krauss (September 2, 2014); BP, *BP announces settlement with Moex/Mitsui of claims between the companies related to the Deepwater Horizon accident*, (May 20, 2011); Marsh, *The 100 Largest Losses 1974-2013, 23rd Ed,* (2014); Washington Post, *Cameron International settles with BP on 2010 oil spill claims*, by Steve Mufson (December 16, 2011); Reuters, *Weatherford to pay $75 million to BP for oil spill*, by Sarah Young (June 21, 2011)

14. Newton.Org, *On the shoulders of Giants*, *available at* www.isaacnewton.org.uk/essays/Giants
15. National Commission on the BP Deepwater Horizon Oil Spill and Offshore Drilling, *Macondo The Gulf Oil Disaster*, Chief Counsel's Report, U.S. Government Publishing Office (2011) *available at* permanent.access.gpo.gov/gpo4390/C21462-407CCRforPrint0.pdf; National Commission on the BP Deepwater Horizon Oil Spill and Offshore Drilling, *Deep Water - The Gulf Oil Disaster and the Future of Offshore Drilling*, Report to the President, U.S. Government Publishing Office (January 2011) *available at* www.gpo.gov/fdsys/pkg/GPO-OILCOMMISSION/pdf/GPO-OILCOMMISSION.pdf; BP, *Deepwater Horizon Accident Investigation Report* (September 8, 2010) *available at* www.bp.com/content/dam/bp /pdf/sustainability/issue-reports/Deepwater_Horizon_Accident_Investigation_Report.pdf; United States Coast Guard, *Report of Investigation into the Circumstances Surrounding the Explosion, Fire, Sinking and Loss of Eleven*

Crew Members, Deepwater Horizon, (September 9, 2011) *available at* www.uscg.mil/hq/ cg5/cg545/dw/exhib/DWH%20ROI%20-%20USCG%20-%20April%2022,%202011.pdf; United States Coast Guard, *Deepwater Horizon Exhibits and Transcripts*, (2010 – 2011) *available at* www.uscg.mil/hq/cg5/cg545/dw/exhib/; CSPAN, *Deepwater Horizon Incident Joint Investigation Testimony*, (2010) *available at* www.c-span.org; *United States of America v. BP Exploration & Production, Inc., et al.*, 21 F.Supp.3d 657 (E.D. Louisiana, Sept. 4, 2014).; MDL 2179 Trial Docs, Phase One, Phase Two, Phase Three, *available at* www.mdl2179trialdocs.com; Expert Report of Greg Childs: Blowout Preventer, *In Re: Oil Spill by the Oil Rid Deepwater Horizon in the Gulf of Mexico on April 20, 2010* (E.D. Louisiana, Sept. 23, 2011); U.S. Chemical Safety and Hazard Investigation Board, *CSB-Final Report BOP*, Engineering Services, (June 2, 2014); Det Norske Veritas, Final Report for the United States Department of the Interior, *Forensic Examination of Deepwater Horizon Blowout Preventer*, (March 20, 2011); Committee on the Analysis on Causes of the Deepwater Horizon Explosion, Fire, and Oil Spill to Identify Measure to Prevent Similar Accident in the Future, *Macondo Well Deepwater Horizon Blowout*, National Academy of Engineering and Nation Research Council (2012)

16. See *supra*, note 15.
17. National Commission on the BP Deepwater Horizon Oil Spill and Offshore Drilling, *Macondo The Gulf Oil Disaster*, Chief Counsel's Report, U.S. Government Publishing Office, at p. 64 (2011), *available at* permanent.access. gpo.gov/gpo4390/C21462-407CCRforPrint0.pdf
18. See *supra*, note 17, at p. 62
19. See *supra*, note 17, at p. 78
20. See *supra*, note 18.
21. See *supra*, note 18.
22. See *supra*, note 17, at p. 63
23. See *supra*, note 17.
24. See *supra*, note 17, p. 98
25. See *supra*, note 17, p. 102
26. See *supra*, note 17, p. 99 and 100
27. Weatherford, *The Importance of Centralization to Well Integrity*, presenter David Westgard (March 16, 2016)
28. See *supra*, note 17, p. 107
29. See *supra*, note 17, p. 230
30. See *supra*, note 17, p. 159 and 160

31. George R.R. Martin, *A Dance with Dragons*, Bantam (October 29, 2013)
32. National Commission on the BP Deepwater Horizon Oil Spill and Offshore Drilling, *Deep Water - The Gulf Oil Disaster and the Future of Offshore Drilling*, Report to the President, U.S. Government Publishing Office at p. 109 (January 2011) *available at* www.gpo.gov/fdsys/pkg/GPO-OILCOMMISSION/pdf/GPO-OILCOMMISSION.pdf
33. See *supra*, note 32.
34. See *supra*, note 32, p. 110
35. See *supra*, note 32, p. 109 and 110
36. See *supra*, note 17, p. 167
37. See *supra*, note 17, p. 168
38. Neal Adams, *Well Control Problems and Solutions*, Petroleum Publishing Company, at p. 110 (1980)
39. See *supra*, note 17, p. 40 and 41
40. See *supra*, note 17, p. 41
41. National Aeronautics and Space Administration, NASA History Office, *Explosion of the Space Shuttle Challenger Address to the Nation*, by President Ronald W. Reagan, (January 28, 1986)
42. Committee on the Analysis on Causes of the Deepwater Horizon Explosion, Fire, and Oil Spill to Identify Measure to Prevent Similar Accident in the Future, *Macondo Well Deepwater Horizon Blowout*, National Academy of Engineering and Nation Research Council at p. 80 (2012)
43. United States District Court Eastern District of Louisiana, In RE: Oil Spill by the Oil Rig Deepwater Horizon in the Gulf of Mexico on April 20, 2010, Day1, *Afternoon Session Transcript of Nonjury Trial Before the Honorable Carl J. Barbier United States District Judge*, (Feb 25, 2013)
44. Wikiquote, *Talk: Albert Einstien*, *available at* en.wikiquote.org/wiki/Talk:Albert_Einstein
45. Superior Energy Services, *HSEQ Policy Statement*, Dave Dunlap *available at* www.superiorenergy.com/pdf/superior-energy-services/hseq-policy-statement.pdf
46. US Well Services, *CEO Commitment*, Health Safety and Environment, *available at* http://uswellservices.com/hse.html
47. United States Department of Labor, OSHA, Safety and Health Topics, *Hydrogen Sulfide – Hazards*, *available at* www.osha.gov
48. Michael Keane, *George S. Patton: Blood, Guts, and Prayer*, Regnery Publishing at p. 153 (2012)
49. Chesapeake, Health and Safety, *Chesapeake Energy Stop Work Authority card*,

available at www.chk.com/Documents/responsibility/StopWorkAuthority Card2016.pdf

50. The King Center, *The Measure of Man*, *available at* www.thekingcenter.org /node/554
51. Dictionary of Quotes, *Claire Nuer – Rock*, from Rebel Thriver, *available at* www.dictionary-quotes.com/claire-nuer-rock
52. The United States Department of Justice, Justice News, *Former Saltwater Disposal Well Operator Indicted in North Dakota on Multiple Felony Charges*, (August 24, 2015)
53. See *supra*, note 52.
54. Wikiquote, *W. Clement Stone Quotes*, *available at* en.wikiquote.org/wiki/ W._Clement_Stone
55. See *supra*, note 52.
56. The United States Department of Justice, Justice News, *Saltwater Disposal Well Operator Pleads Guilty to Multiple Felony Charges in Connection with Operation of Well*, (September 26, 2014)
57. United States District Court for the District of North Dakota Southwestern Division, *United States of American v. Man #1*, Indictment, (August 6, 2015), *available at* www.cnsenvironmentallaw.com
58. See *supra*, note 57.
59. GoodReads, *Mahatma Gandhi Quotes*, *available at* www.goodreads.com/quo tes/30431-earth-provides-enough-to-satisfy-every-man-s-needs-but-not
60. Aubrey McClendon, *American Energy Partners communication*, (2015)
61. Mike Cochran, *Claytie The Roller-Coaster Life of A Texas Wildcatter*, Texas A&M University Press (2007)
62. Wikiquote, *Work*, Thomas Edison as quoted in Ford Times, Vol. 6 (1912) *available at* en.wikiquote.org/wiki/Work
63. Wikiquote, *George Bernard Shaw*, there is uncertainty on the attribution *available at* en.wikiquote.org/wiki/ George_Bernard_Shaw
64. OSHA, *Fatality Narrative*, Accident 200980290, Inspection 303668537, (2002); Ray Ring, High Country News, Fatalities in the energy fields: 2000- 2006, April 2, 2007, *available at* www.hcn.org/issues/343/16931
65. See *supra*, note 64.
66. ExxonMobil, Speeches, *Advancing a Culture of Safety*, by Rex W. Tillerson (November 3, 2015) *available at* corporate.exxonmobil.com/en/ company/news-and-updates/speeches/advancing-a-culture-of-safety
67. United States Department of Justice, The United States Attorney's Office, Eastern District of Louisiana, News, *Oil Company Sentenced for Multiple*

Felonies Related to Violations of Offshore Oil Production Safety and Environmental Regulation (April 6, 2016)

68. G. Rivlin, *Understanding the Law*, Oxford University Press at p. 47 (2012)
69. See supra, note 67.
70. See supra, note 67.
71. Tryon Edwards, *A Dictionary of Thoughts*, Cassell Pub. at p. 486 (1899)
72. Francis Bacon, *Of Studies*, Essay (1625)
73. United States Department of Labor, Bureau of Labor Statistics, *Fatal injuries in the U.S. workplace rise in 2014* (April 28, 2016) *available at* www.bls.gov/opub/ted/2016/fatal-injuries-in-us-workplaces-rise-in-2014.htm
74. Supreme Court of Oklahoma, *Plaintiffs/Appellees/Petitioners v. Defendant/Appellant/Respondent*, No. 101686, 229 P.3d 540 (2010) 2010 OK 10 (Feb, 9, 2010)
75. See *supra*, note 74.
76. Daniel J. Boorstin, *The Discoverers*, Vintage Books, (February 1985)
77. USDOL, Bureau of Labor Statistics, *Databases, Tables, & Calculators by Subject – Workplace Injuries*, data search analysis and tables analysis includes NAICS codes 211, 213111, 213112. 2014 BLS data considered preliminary at the time of research, *available at* ww.bls.gov/data/#injuries
78. See *supra*, note 77.
79. See *supra*, note 77.
80. USDOL, OSHA, Safety and Health Topics, Oil and Gas Extraction, *Safety Hazards Associated with Oil and Gas Extraction Activities*, *available at* www.osha.gov/SLTC/oilgaswelldrilling/safety hazards.html
81. See *supra*, note 77.
82. Association for Safe International Road Travel (ASIRT), *Road Crash Statistics*, *available at* asirt.org/initiatives/informing-road-users/road-safety-facts/road-crash-statistics
83. GoodReads, *Van Gogh*, Quotable Quotes, *available at* www.goodreads.com/quotes/638134-if-you-don-t-have-a-dog--at-least-one--there-is-not
84. See *supra*, note 82
85. NHTSA, *Traffic Safety Facts 2014 Data* (April 2016) *available at* crashstats.nhtsa.dot.gov/Api/Public/Publication/812265
86. See *supra*, note 85.
87. AAA Foundation for Traffic Safety, *Prevalence of Motor Vehicle Crashes Involving Drowsy Drivers, United States, 2009 – 2013* (November 2014) *available at* www.aaafoundation.org/sites/default/files/AAAFoundation-DrowsyDriving-Nov2014.pdf

88. Centers for Disease Control and Prevention, Injury Prevention & Control: Motor Vehicle Safety, *Impaired Driving: Get the Facts*, *available at* www.cdc.gov/motorvehiclesafety/impaired_driving/impaired-drv_factsheet.html
89. Distraction.gov, *Traffic Safety Facts – Distracted Driving 2013*, U.S.DOT, NHTSA, (April 2015) *available at* www.distraction.gov/downloads/pdfs/Distracted_Driv ing_2013_Research_note.pdf
90. See *supra*, note 82.
91. Centers for Disease Control and Prevention, *Seat Belts: Get the Facts*, *available at* www.cdc.gov/motorvehiclesafety/seatbelts/facts.html
92. See *supra*, note 82.
93. Oil & Gas Safety and Health Conference 2014 OSHA Exploration & Production, *Driving Safety for the Oilfield – Panel Discussion*, presentation by Kyla Retzer, John Stephens, Anthony Zacniewski
94. See *supra*, note 93.
95. See *supra*, note 93.
96. Texas A&M University, *Alumnus: Tillman talks leadership*, Summer Special 2015 Newsletter (2015) *available at* engineering.tamu.edu/media/2750940/texas-am-chemical-engineering-summer-special-2015-e-newsletter.pdf
97. Supreme Court of Mississippi, *Oil Company and Company Man v. Administrator of the Estate of Deceased; and Wife*, No. 97-CA-01447-SCT, 829 So.2d 1 (2002)
98. Huffington Post, *It's Not What You Look at That Matters, It's What You See*, by Dennis Merrit Jones (June 3, 2013)
99. Upton Sinclair, *Oil!*, Albert and Charles Boni, Inc. (1927), Penguin Books at p. 38 (2007)
100. Railroad Commission of Texas, Oil & Gas, Compliance & Enforcement, *Blowouts and Well Control Problems*, (data pulled in 2015), *available* at www.rrc.state.tx.us/oil-gas/compliance-enforcement/blowouts-and-well-control-problems/
101. Philip Singerman, *Red Adair: An American Hero*, Bloomsbury Publishing at p. 104 (June 25,1990)
102. The New York Times, *Coots Matthews, Cantankerous Hellfighter, Dies at 86*, by Douglas Martin (April 7, 2010)
103. W.C. Goins, *Blowout Prevention*, Gulf Publishing Company at p. 51 (1969)
104. United States District Court Eastern District of Louisiana, In RE: Oil Spill by the Oil Rig Deepwater Horizon in the Gulf of Mexico on April 20,

2010, Docket 10-CV-02771, In Re: The Complaint and Petition of Triton Asset Leasing GmbH, et al, Docket 10-CV-4536, United States of America v. BP Exploration & Production, Inc. et al., *Day 1 Afternoon Session Transcript of Nonjury Trial Before the Honorable Carl J. Barbier United States District Judge*, (February 25, 2013) *available at* www.mdl2179trialdocs.com/releases/release201302250700001/2013-02-25_BP_Trial_Day_01_PM-Final.pdf

105. See *supra*, note 104.
106. See *supra*, note 104.
107. See *supra*, note 101, p. 21
108. See *supra*, note 104.
109. See *supra*, note 104.
110. See *supra*, note 104.
111. See *supra*, note 104.
112. Sun Tzu, *The Art of War*, Translation, Essays and Commentary by the Denma Translation Group, Shambhala p. 50 (2001)

Chapter Three

1. Whitney Tilson, *Whitney Tilson's 2007 Wesco Annual Meeting Notes,* Q&A period with Charlie Munger (May 9, 2007) *available at* www.valuewalk.com
2. NASA, *Apollo 16 Cuff Checklists – Drill Core Sample*, *available at* www.hq.nasa.gov/alsj/a16/a16.lmpev119.gif
3. NASA, *Apollo 16 Cuff Checklist – Core Extraction and GeoPrep*, *available at* www.hq.nasa.gov/alsj/a16/a16.lmpev120.gif
4. AZ Quotes, *Burt Rutan*, *available at* www.azquotes.com/quote/1494236
5. Patrick O'Brian, *Pablo Ruiz Picasso*, William Collin Sons & Co. (1976)
6. Brian Tracy, *Time Management*, American Management Association at p. 29 and 30 (2013) *available at* www.amanet.org/time_mgmnt_mini.pdf
7. Getty Store, *Magnet - Getty Quote – Formula for Success*, J. Paul Getty *available at* shop.getty.edu/products/magnet-getty-quote-formula-for-success
8. L.P. Dake, *The Practice of Reservoir Engineering*, Elsevier at p. 10 (2001)

Chapter Four

1. Rigzone, *Baker Hughes CEO: Turn Adversity into Opportunity*, by Valerie Jones, (May 7, 2015) *available at* www.rigzone.com/news/oil_gas/a/138481 /Baker_Hughes_CEO_Turn_Adversity_into_Opportunity
2. New York Times, On This Day, *Mies van der Rohe Dies at 83; Leader of Modern Architecture*, (August 19,1969) *available at* www.nytimes.com/learning/general/onthisday/bday/0327.html

3. OSHA, *Summary of Fatality Investigation*, Accident 200783454 , Inspection 307489161 (2005)
4. See *supra,* note 3.
5. See *supra,* note 3.
6. Byron Davenport, *Handbook of Drilling Practices*, Gulf Publishing Company at p. 9 in Preface chapter (1984)
7. Robert Hazen, *The Story of Earth: The First 4.5 Billion Years*, from Stardust to Living Planet, Penguin Books at p. 13 (2013)
8. Jacqueline Sweeney, *Incredible Quotations*, Scholastic Inc. at p.26 (1997)
9. Edwin Lefèvre, *Reminiscences of a Stock Operator*, Wiley at p. 82 (1923)
10. Columbia University, *Pablo Picasso – Statement to Marius de Zayas*, (1923) *available at* www.learn.columbia.edu/monographs/picmon/pdf/art_hum_reading_49.pdf
11. GoodReads, *Plato – Quotable Quote*, *available at* www.goodreads.com /quotes/20382-those-who-tell-the-stories-rule-society
12. Ronald Reagan, *Farewell Address to the Nation*, (January 11, 1989) *available at* www.reaganlibrary.archives.gov/archives/speeches/1989 /011189i.htm
13. Daniel Coenn, *Confucius: His Words*, BookRix, (June 8, 2014)
14. OSHA, *Safety Narrative*, Accident 200631240, Inspection 307070516
15. OSHA, *Safety Narrative*, Accident 200631265, Inspection 307073395
16. Wikiquote, *Niccolò Machiavelli*, *available at* en.wikiquote.org/wiki/Niccol%C3%B2_Machiavelli
17. OSHA, *Safety Narrative*, Accident 201573409, Inspection 310472410 (2008)
18. OSHA, *Accident Report*, Accident 200556652, Inspection 314772500 (2011)
19. Readers Digest, *Quotable Quotes*, Reader's Digest Association (1997)
20. University at Albany, *Annual Symposium on Information Assurance – Security Notes*, *available at* www.albany.edu/iasymposium/awards.2014.shtml
21. Caltech, *Citizenship in a Republic*, by Theodore Roosevelt (April 10, 1910) *available at* design.caltech.edu/erik/Misc/Citizenship_in_a_Republic.pdf
22. Library of Economics and Liberty, *On Liberty*, by John Stuart Mill, (1869) *available at* www.econlib.org/library/Mill/mlLbty1.html
23. W.C. Goins, *Blowout Prevention*, Gulf Publishing Company at p. 1 (1969)
24. Miguel De Cervantes, *Don Quixote*, translated by Edith Grossman, Harper Perennial at p. 734 (April 26, 2005)
25. 2014 Neonatal Symposium, *Clinical Research*, (February 7, 2014)
26. OSHA, *Inspection Report*, RID 0830500, Inspection 580518 (2012)
27. See *supra*, note 26.

28. AZ Quotes, *Cus D'Amato, available at* www.azquotes.com/quote/891270
29. OSHA, *Investigation Summary*, Accident 200213056, Inspection 316283118
30. OSHA, *Safety Narrative*, Accident 918755, Inspection 311323075 (2007)
31. Popular Science Monthly, *Freak Accidents*, by Grahame at p. 42 (Nov.1931)

Chapter Five

1. Twitter, *Cornelius Fichtner Tweet*, (July 7, 2009) *available at* twitter.com/corneliusficht/status/2515541824
2. T. Boone Pickens, *The First Billion Is the Hardest*, Crown Business at p. 125 and 127 (2009)
3. Ranker, *The Most Famous Confucius Quotes*, *available at* www.ranker. com /list/a-list-of-famous-confucius-quotes/reference
4. Rick Bass, *Oil Notes*, Houghton Mifflin at p. 13 (1989)
5. Halliburton, *Halliburton Life Rules*, presentation, (2013) *available at* www.halliburton.com/public/pubsdata/hse/HAL_Life_Rules_ver8_LC.pptx

Chapter Six

1. Biblehub, *New Living Translation*, Tyndale House Publishers, Inc. (2007) available *at* biblehub.com/nlt/job/38.htm
2. NASA, *How Old is the Universe?, available at* map.gsfc.nasa.gov/universe/uni_age.html; Stanford, *How old is the Sun?, available at* solar-center.stanford.edu/FAQ/Qage.html; NASA, *What Is Earth? available at* www.nasa.gov/audience/forstudents/5-8/features/nasa-knows/what-is-earth-58.html; Gulfport, *Operations Overview, available at* www.gulfport energy.com/operations/Utica; AOGR, *Marcellus Shale Play's Vast Resource Potential Creating Stir In Appalachia*, by Terry Engelder and Gary G. Lash, *available at* www.aogr.com/magazine/cover-story/marcellus-shale-plays-vast-resource-potential-creating-stir-in-appalachia; Continental, *Anadarko Woodford: The SCOOP, available at* media.corporate-ir.net/media_files /irol/19/197380/CLR_Sunday_10_7_12_Anadarko_Woodford_3.pdf; EERC, *Beyond the Boom, available at* www.undeerc.org/bakken/bakken formation.aspx; Kansas Geological Survey, Public Information Circular (PIC) 33, *The Mississippian Limestone Play in Kansas: Oil and Gas in a Complex Geologic Setting, available at* www.kgs.ku.edu/Publications /PIC/pic33.html; NETL, *Modern Shale Gas Development in the United States: An Update*, (2013); University of California, *The Permian Period, available at* www.ucmp. berkeley.edu/permian/permian.php; The Oil & Gas Year, *Abu Dhabi 2010*, Wildcat Publishing; IPAA, *Haynesville: Natural Gas Comeback in Texas & Louisiana, available at* oilindependents.org/haynesville-natural-gas-

comeback-in-texas-louisiana; Omics International, *Eagle Ford Formation, available* at research.omicsgroup.org/index.php/Eagle_Ford_Formation; Bay Nature, *How the Monterey Shale came to be,* by Sarah Phelan (September 2, 2013) *available at* baynature.org/article/how-the-monterey-shale-came-to-be/; Boeda E., Bonilauri S., Connan J., Jarvie D., Mercier N.,Tobey M., Valladas H. & Al Sakhel H., *New Evidence for Significant Use of Bitumen in Middle Palaeolithic Technical Systems at Umm el Tlel (Syria) around 70,000 BP,* Paléorient (2008) *available at* www.persee.fr/doc/paleo_0153-9345_2008_num_34_2_5257; Aramco World, *Bitumen – A History*, (1984) available at archive.aramcoworld.com/issue/198406/bitumen.-.a.history.htm

3. National Safety Council (NSC), *NSC Poll: One-third of American Workers Do Not Feel Prepared for an Emergency*, (September 30, 2016) *available at* www.n sc. org/Connect/NSCNewsReleases/Lists/Posts/Post.aspx?ID=142
4. Philip Singerman, *Red Adair: An American Hero*, Bloomsbury Publishing at p. 22 (June 25,1990)
5. Penn Medicine, Lancaster General Health, *The golden hour in trauma: Dogma or medical folklore?*, Elsevier -Injury (2016) *available at* www.lancastergeneral health.org
6. ASBMB, *Chance Favors the Prepared Mind*, by Shawn Drew (March 2010) *available at* www.asbmb.org/
7. Jones Energy, *2015 Annual Report and Form 10-K Proxy Statement*, (2015) *available at* investors.jonesenergy.com/investors/#AnnualReports
8. Court of Appeals of Texas, First District, Houston, *Appellant, v. Appellees*, No. 01-11-00079-CV (January 31, 2013)
9. GoodReads, *Ralph Waldo Emerson*, Quotable Quote, *available at* www.goodreads.com/quotes/57008-shallow-men-believe-in-luck-or-in-circumstance-strong-men
10. OSHA, *Inspection Narrative*, Accident 200758878, Inspection 313151029
11. See *supra*, note 10.
12. Daniel Ebner, *Formal and Informal Strategic Planning*, S. Gabler at p. 4 (2014)
13. Andrew Grove, *Only the Paranoid Survive*, Crown Business (March 16, 1999)
14. OSHA, *Safety Narrative*, Complaint 208205468 , Inspection 314663188
15. W.H. Auden and Louis Kronenberger, *The Viking Book of Aphorisms*, Barnes and Noble Books at p. 99 (1962)
16. United States District Court, S.D. Texas Houston Division, *Plaintiff v. Defendants*, Civil Action No. H-13-2844, (July 7, 2014)
17. James W. Kinnear, *The Man Who Wore The Star*, The Business Council of New York State at p. 3 (2001)

18. Wikiquote, *Murphy's law Quotes, available at* en.wikiquote.org/wiki/Murphy's_law
19. William Gibson, *Zero History*, G.P. Putnam's Sons at p. 276 (Sept. 7, 2010)
20. United States District Court, N.D., Texas, Wichita Falls Division, *Plaintiffs, v. Defendants*, Civil Action No. 7:12-cv-00133-O. (2013 - 2014)
21. Rogers, *Illiterate Digest*, Kessinger Publishing at p. 217-218 (Mar 31, 2003)
22. Sun Tzu, *The Art of War*, Value Classic Reprints at p. 5 (Sept. 22, 2016)
23. U.S. Chemical Safety and Hazard Investigation Board (CSB), *Incident Investigation Report*, Gas Well Explosion, Bienville Parish, LA (July 5, 1999); OSHA, *Inspection Detail*, Report ID 0625700 Inspection 301994315 (April 22, 1999)
24. Environmental Resource Center, *OSHA Proposes $208,750 Penalty Against 3 Companies For Safety Violations Resulting in Seven Fatalities*, (May 3, 1999)
25. See *supra*, note 23.
26. See *supra*, note 23.
27. See *supra*, note 23.
28. J.R.R. Tolkien, *The Hobbit*, H. Mifflin Harcourt at p. 235 (Sept. 18, 2012)
29. RigZone, *Baker Hughes CEO: Turn Adversity into Opportunity*, Valerie Jones (May 7, 2015) *available at* www.rigzone.com/news/oil_gas/a/138481/Baker_Hughes_CEO_Turn_Adversity_into_Opportunity
30. Joseph P. Zbilut and Alessandro Giuliana, *The Latent Order of Complexity*, Nova Science Publishers at p. 1(2008)

Chapter Seven

1. W.J. Rock, *Handbook for Civilization*, Xlibris Corporation at p. 164 (2010)
2. Business Insider, *This Steve Jobs quote perfectly sums up the difference between billionaires and the rest of us*, Kathleen Elkins (January 27, 2016) *available at* www.businessinsider.com/steve-jobs-difference-between-billionaires-and-the-rest-of-us-2016-1
3. Philip Gooden, *Skyscrapers, Hemlines and the Eddie Murphy Rule: Life's Hidden Laws, Rules & Theories*, Bloomsbury (2015)

Chapter Eight

1. Rockefeller Achieve Center, *Home Page Quote*, *available at* rockarch.org
2. Rick Bass, *Oil Notes*, Houghton Mifflin at p. 30 (1989)
3. Ari Ben-Menahem, *Historical Encyclopedia of Natural and Mathematical Sciences*, Volume 1, Springer at p. 341 (2009)
4. Lee Iacocca, *Iacocca An Autobiography*, Bantam Books at p. 357 (1984)

5. National Safety Council, *The Journey to Safety Excellence*, (2016) *available at* www.nsc.org/JSEWorkplaceDocuments/JSE-Infographic-Printable.PDF
6. OSHA, *Accident Report*, Accident 202470431, Inspection 314824681 (2011)
7. OSHA, *Safety Narrative*, Accident 200541985, Inspection 314108119 (2010)
8. USGS, *Things are not always what they seem; the first appearance deceives many?*, Hawaiian Volcano Observatory (February 24, 2011) *available at* hvo.wr.usgs.gov/volcanowatch/archive/2011/11_02_24.html
9. Space.com, *Can Life on Earth Escape the Swelling Sun?,* Jeremy Hsu, Astrobiology Magazine (August 3, 2009) *available at* www.space.com /7084-life-earth-escape-swelling-sun.html
10. BOEM, *Investigation of Loss of Well Control*, OCS-G 02925, MMS 2008-054, Gulf of Mexico Off the Louisiana Coast, U.S. Department of the Interior, Minerals Management Service (MMS), New Orleans (November 2008)
11. James Cumming, *Resolves, Divine, Moral, and Political of Owen Felltham*, John Hatchard and Son at p. 308 (1820)
12. See *supra*, note 10.
13. Walter Scott, *Harold the Dauntless*, J. Ballantyne & Co. at p. 12-13 (1817)
14. OSHA, *Safety Narrative*, Accident 200542033, Inspection 314107467 (2010)
15. See *supra*, note 14.
16. See *supra*, note 14.
17. See *supra*, note 14.
18. See *supra*, note 14.
19. See *supra*, note 14.
20. See *supra*, note 14.
21. Readers Digest, *Quotable Quotes*, Reader's Digest Association (1997)
22. United States District Court, W.D. Oklahoma, *Plaintiffs, v. Defendant/Third Party Plaintiff, v. Third-Party Defendant*, Case No. CIV-13-118-M (December 16, 2014)
23. Wikiquote, *Talk: Nikola Tesla, available at* en.wikiquote.org/wiki/Talk: Nikola_Tesla
24. The Economist, *An interview with George Mitchell*, Schumpeter Business and Management (August 1, 2013) *available at* www.economist.com/ blogs/ schumpeter/2013/08/interview-george-mitchell
25. The Hon. Lord Cullen, *The Public Inquiry into the Piper Alpha Disaster*, The Department of Energy. Volume 1 and 2, H.M.S.O. (November 1990); NASA Safety Center (NSC), *System Failure Case Study Details- The Case for Safety, The North Sea Piper Alpha Disaster* (May 6, 2013) Case Study and Presentation *available at* nsc.nasa.gov/SFCS/SystemFailureCaseStudy

/Details/112; Risk Analysis, *Learning from the Piper Alpha Accident: A Postmortem, Analysis of Technical and Organizational Factors*, by Elisabeth Pate-Cornell, Volume 13, No.2 (1993) *available at* www.researchgate.net; Scottish Courts and Tribunals, *Appendix to the Opinions of the Judges in Reclaiming Motions in the Causes*, Pursuers and Reclaimers Against Defenders and Respondents, *available at* www.scotcourts.gov.uk/; Herald Scotland, *Piper Alpha valve fitter given a notice of blame* (April 21, 1989) *available at* www.heraldscotland.com/news/11896584.Piper_Alpha_valve_fitter_given_a_notice_of_blame; Brian Appleton, *Piper Alpha Appleton BBC Lecture*, *available at* www.youtube.com/watch?v=S9h8MKG88_U

26. See *supra*, note 25.
27. Carl Sagan, *Cosmos*, Ballantine Books at p. 295 -296 (December 10, 2013)
28. See *supra*, note 25
29. John F. Kennedy, *Remarks at the America's Cup Dinner*, (September 14, 1962) *available at* hwww.jfklibrary .org/Research/Research-Aids/JFK-Speeches/Americas-Cup-Dinner_19620914.aspx
30. See *supra*, note 25.
31. Stephan McGinty, *Fire In The Night*, Macmillan at p. 102 (2008)
32. Bliss Carman, et al. eds, *The World's Best Poetry*, Volume VII, Horatius at the Bridge, Thomas Babington, Lord Macaulay (1904)
33. Brian Appleton, *Piper Alpha Appleton BBC Lecture*, *available at* www.youtube.com/watch?v=S9h8MKG88_U
34. Dave Hager, *Workplace Safety at Devon*, *available at* www.devonenergy.com /social-responsibility/health-safety

Chapter Nine

1. Mike Berardino, *Mike Tyson explains one of his most famous quotes*, Sun Sentinel, (November 9, 2012) *available at* articles.sun-sentinel.com/2012-11-09/sports/sfl-mike-tyson-explains-one-of-his-most-famous-quotes-20121109_1_mike-tyson-undisputed-truth-famous-quotes
2. United States Court of Appeals, *Fifth Circuit, United States of America, Plaintiff-Appellant v. Defendants-Appellees*, No. 14-30122 (March 11, 2015)
3. Queen Rania Al Abdullah, *Queen Rania to Oprah: "I think what would surprise most people is just how alike we are"*, Media Center, Rania (May 17, 2006) *available at* www.queenrania.jo/en
4. See *supra*, note 2.

Chapter Ten

1. Reference, *How many people does the average person physically meet in a lifetime?*, *available at* www.reference.com/world-view/many-people-average-person-

physically-meet-lifetime-72cdc2307255db8e; Quora, *How many people does the average person meet over the course of a lifetime?*, by Shoaib Deoda, *available at* www.quora.com/How-many-people-does-the-average-person-meet-over-the-course-of-a-lifetime; Funders and Founders, *Why We Live – Counting The People Your Life Impacts*, by Anna Vital (April 29, 2013) *available at* fundersandfounders.com/counting-the-people-you-impact/

2. Upton Sinclair, *Oil!*, Albert and Charles Boni, Inc. (1927), Penguin Books at p. 548(2007)
3. Schlumberger, *Schorn Speaks at Barclays CEO Energy-Power Conference*, Patrick Schorn (September 7, 2016) *available at* www.slb.com/news/presentations /2016/2016_0907_schorn_barclays.aspx
4. Philip Singerman, *Red Adair: An American Hero*, Bloomsbury Publishing (June 25,1990)
5. Wall Street Journal, *After Tough Year, Hess CEO Remains Focused*, by Tom Fowler (February 13, 2014) *available at* www.wsj.com
6. PWC, *CEO Interview: John J. Christmann IV*, US CEO Survey (Fall 2015) *available at* www.pwc.com/us/en/ceo-survey/ceo-interviews.html
7. Harold Hamm, *Harold Hamm – Oklahoma Christian Q&A*, Oklahoma Christian, *available at* www.youtube.com/watch?v=bcQyLr4wLhM
8. Motley Fool, *National Oilwell Varco CEO Clay Williams Sits Down With The Motley Fool*, by Tyler Crowe (Jan. 23, 2016)
9. Daniel Patrick O'Brien, *Business Measurements for Safety Performance*, Lewis Publishers at p. 39 (2000)
10. Court of Appeal of Louisiana, Third Circuit, *Plaintiffs-Appellees, v. Defendants-Appellants* , No. 83-522 (1984)
11. Diane Beck, *Project Management Secrets*, Bush Street Press at p. 87 (2013)

About the Author

Matthew J. Hatami is a petroleum engineer and entrepreneur. He holds a degree in Petroleum and Natural Gas Engineering from West Virginia University and an MBA from Columbia University. He is a licensed Professional Petroleum Engineer in the State of Oklahoma. He started his career 16 years ago, working as a field engineer in Hobbs, New Mexico. He has worked in oilfield operations across multiple basins in the United States and overseas, primarily focusing on shale development and frontier shale exploration. He has worked on the oilfield services side of the industry and on the operating side, including positions working with geology, land, drilling, completions, production, reservoir, regulatory, legal, accounting, finance, and corporate strategy. Before venturing into the entrepreneurial arena, he was the Director of Resource Development for American Energy Partners and American Energy Global Partners, with the latter focused on opportunities to create value by employing American shale technology in international markets. He lives with his wife in Oklahoma City, Oklahoma.

Made in the USA
Coppell, TX
13 March 2025

47003931R00194